EVOLUTION

BY MEANS OF HYBRIDIZATION

EVOLUTION

BY MEANS OF HYBRIDIZATION

BY

J. P. LOTSY

THE HAGUE
MARTINUS NIJHOFF
1916

ISBN 978-94-011-8381-9   ISBN 978-94-011-9072-5 (eBook)
DOI 10.1007/978-94-011-9072-5

Softcover reprint of the hardcover 1st edition 1916

To the Memory of Charles Darwin, this Sketch is dedicated by the Author as a Tribute to his Character and to his Work which lead to the general Recognition of the Principle of Evolution.

PREFACE.

As a Preface is in reality a Postscriptum, the author may be permitted to open it by mentioning omissions.

The chief sin of omission he committed, is evidently the insufficient justice he did to the writings of Anton Kerner von Marilaun, who was — he wants to state this *explicitly* — the first to recognize fully the significance of crossing as the underlying cause of the origin of species.

What else should a preface say?

If the work is as condensed as the present one, it may perhaps suffise to repeat what Linnaeus said to Haller:

Si quos in me vidisti errores, Tu sapientior haec ignoscas.... Quos plures apud me detegere potes, eo gratior ero, tum possum omnia corrigere vivus; post mortem non licet emendare propria opuscula.

By which however the author does not consider himself bound to gratefulness for *every* kind of criticism.

He is f. i. very little impressed by the kind of criticism which calls it „inconceivable" „verging on the absurd" etc., to believe that crossing can ever have been the underlying cause of the origin of new species, from authors who firmly believe that the origin of new species should be ascribed to some kind of variability; because it seems to him „absurd" that those who advocate the origin of new species from *a single* ancestral one, should reproach an author who defends such an origin from *two* ancestral species, of stating an „inconceivable" opinion.

To save another kind of critics unnecessary trouble, the author is fully prepared to admit unhesitatingly that his theory explains but part of the problem of evolution, so that there is ample room for them to take part in the exploration of this most interesting field of investigation; but he may be permitted to remark that the rest, left unexplained by theories based on variability, is not smaller than that left unexplained by his theory.

The advantage of the theory, here sketched out, seems to the author to be, that it is based on the experimentally proven fact that crossing gives rise to new forms, while, in as much as the real existence of hereditary variability is not proven, all theories based on variability have a less solid basis.

Unfortunately the final basis of evolution — the constitution of the germplasma — remains as yet an utter mystery, so that we know next to nothing of the final cause of evolution; the more reason to keep facts and mere surmises well separated. This the author has aimed at, and he flatters himself that, by so doing, he has cut a small peep-hole in the curtain which hides the stage on which evolution takes its course. Criticism within *this* contention, e. g. dealing with the part of evolution which is discussed in this sketch, will indeed be most welcome to him and will be considered in a careful manner and a grateful spirit. Du choc des opinions jaillit la vérité!

HAARLEM, February 16th 1916.

PART I.

INTRODUCTION.

WHAT IS EVOLUTION?

From the very moment that mankind began to reflect, the desire was born to know how all those different objects which man sees around him, came into being and many are the opinions, vague and detailed ones, which he has held and still holds on this momentous question.

To the unexperienced, nature has the aspect of stability, of being the same to day as yesterday and this conviction underlies all legends of deities, creating the earth in a few days, so as to have it ready for the crowning appearance of mankind.

A little experience and a little reflection however, teaches us that the apparent stability of nature is deceptive, and that this impression is caused exclusively by the short time of observation, allowed to each single individual.

A human life lasts but a moment of eternity and even the lives of many successive individuals are insufficient to observe slow changes in the aspect of nature.

Even if a man could return to the earth every few centuries, he would yet be unable to see much change.

A LUTHER, coming back to the Wartburg to day, could easily revisit all spots once familiar to Junker Jörg and believe that but a few days had passed since he last saw them.

The stone which used to serve him as a resting place, is yet there, it is covered by the same species of mosses which grew there in the 16th century; the same kinds of flowers adorn now as then, meadows and rocks, and Landgrafenschlucht, as well as Hohe Sonne, would be familiar landmarks to him, while his feet would carry him, as in former ages, through Drachenschlucht and Annathal, back to the Marienthal from where he could ascend the Wartburg along familiar paths, and once there, even revisit his former abode.

So it would appear, even to a returning Luther, as if the landscape of Thuringia — with the exception of minor alterations made bij man — were a permanent and unchangeable result of the working of a Creator.

And yet this permanency of Nature is a delusion and a snare.

We need not go very far back into the history of our globe, to find on the spot, now occupied by the Thuringian hills, a monotonous flat track of land, and if we recede a little further into time, we see that this continent was preceded by an Archipelago of small isles, while a little further back still, the spot now occupied by Thuringia, was covered by the waters of an ocean, on the bottom of which fragments of ancient rocks were interspersed by a few coral reefs, as a token of the power of the infinitely small, of the action of myriads of minute living beings.

Gradually, slowly, ever so slowly, these fragments of rocks were kitted together by chalky and other substances, a kit including the coral reefs also, and forming a new layer around that one, of the numberless

celestial bodies which we call the globe, a layer destined to persist for countless ages as a horizontal crust.
But below this crust, forces whose action never ceases continued their work of change. Under the influence of the cold of celestial Space, our globe was forced
to part with much of its heat and consequently had to
contract. This caused the appearance of folds on her
surface, of one of which the Thuringian hills are a
rest. By such folding the fragments of rocks, as well
as the coral reefs enclosed in the, up till then horizontal, ocean bottom underwent a considerable upheaval.

And of the fold, so formed, Thuringia is but a *small*
rest; it has been lowered by the action of wind and
water and the latter especially cut into it those deep
valleys of Drachenschlucht and Mariathal which now
delight our eyes, and isolated the Wartburgtop, enclosing in its conglomerate a coral-reef, destined to form
in a distant future the foundation of the Castle which
should serve Luther as a temporary abode.

So it happened, that the translation of the Bible took
place on the rests of small living beings, who countless
ages ago, peopled an ocean, expanded hundreds of feet
below the point where those corals are now found.
What seemed so permanent, was thus but a short phase
in the chain of changes which the earth and all it
contains or supports, has to undergo. These changes of
which there is no escape, neither for rocks nor for seas,
nor for living beings, we call evolution.

No wonder that man, once convinced of the existence of so universal a change, wanted to know *how* evolution acted, and no wonder that numerous different opinions have been held on this fascinating subject. Even yet, we are far from a communis opinio.

We are best informed as to the non-living part of the universe. Ingeniously worked out methods have given us the certainty that all celestial bodies consist chiefly of the same kind of substances, and we have also ascertained that a comparatively small number of primary substances — the so called elements —suffise to compose alle those numerous substances which occur in nature or can be made in our laboratories.

The discovery of the existence of these elements led at first and had to lead, to the conception that elements were absolutely independent primary substances, but PROUT, a London physician (obiit 1850), called already in 1815, attention to the fact that the atomic weights of all elements are invariably integer multiples of the atomic weight of the element Hydrogen, and from this knowledge deducted the hypothesis that all elements might very well be derivatives, by a process of condensation, of Hydrogen.

It has later been shown that the integrity of these multiples is not perfect, but the ascertained deviations were not sufficiently great, to force us to discard Prout's supposition of the existence of some kind of connexion between the different elements.

On the contrary, Mendelejeff's periodical system gave new support to Prout's idea.

This system arranges the elements in a series according to their atomic weights. Part of the elements so arranged, forms a series with pretty equal intervals between the successive atomic weights; at several points however considerable gaps occur. From this irregularity MENDELEJEFF concluded that there must exist other elements — till then unknown — and he ventured to predict that the gaps in his series would gradually be filled up by these missing links, of which he even predicted the atomic weights.

In many cases the gaps have really been thus filled, and this of course gave strong support to the idea that elements are not such independent primary substances, as one had previously supposed them to be.

Yet, decisive proof for a genetic connexion between the several elements lacked until the discovery of the radioactive elements by Monsieur et Madame CURIE, demonstrating the direct transformability of the elements.

We now know that a number of elements, among which Ionium, Radium, Polonium and very probably lead also, can be directly derived from Uranium by a process of splitting off Helium, with the atomic weight 4, and electrons, whose apparent mass is but $1/_{2000}$ part of a hydrogen-atom.

Unfortunately, it is as yet impossible to excert even the slightest influence on the process of radioactivity, and it is especially unfortunate that we cannot reverse the process and thus, by adding Helium and Electrons to lead f. i., build up Uranium.

But the fact that other elements e. g. Potassium and Rubidium also, though in a much slighter degree, show signs of radioactivity, justifies the supposition that Helium takes part in the building up of all elements,

and this is of so great an importance, because the nebulae which are considered to be the first stages in
the evolution of the celestial bodies consist, if not exclusively, at least chiefly of Helium.

This leads to the conception that during the development of celestial bodies from nebulae, the elements
were formed of Helium or of a similar substance, and
that these elements, later on, united for the greater
part with other elements, thus forming more complicated chemical bodies.

Bij this process also, water and rocks were formed
which, under the influence of the changes in the form
of the earth's crust, caused by cooling off, changed
places, so that where mountains once were, oceans now
spread their waters and reversely.

Of course, the time required to bring about all these
changes has been a long one; its length can be determined approximately by studying the behaviour of
radioactive minerals.

Radio-activity namely, proceeds with a velocity,
independent of external circumstances such as heat
etc., and this allows us to calculate, if the quantity
of Helium present in a mineral is known, by determination of the quantity of Helium yearly split off by
this mineral, the minimum age of it. It was thus possible to state positively, that Titanite or Grothite from
the Archaic Period must, at the very least, be 710 million years of age, and this agrees quite well with the
estimate of the earth's age at 1000 millions of years,
derived from data, concerning the formation of the
great saltbeds.

This means, that if the complete history of the earth could be compressed into a course of daily one-hour lectures lasting one and a half year, the life of a student — suppose he could begin attending them immediately after birth and continue to do so up to the completion of his 80th year — would be just sufficient to hear two seconds of these lectures.

Once — or repeatedly — during this long, long period, living beings arose from much more complicated combinations of the elements also.

Keen as this deduction may seem to be, in the face of the fact that generatio aequivoca has never yet been demonstrated, we are nevertheless forced to it.

1stly. because there has been a time in the history of the earth, during which life was impossible upon it.

2dly. because we know that no other elements occur in the living beings than in non-living nature, so that both are composed of the same substances.

3dly. because we know that even at the present time, the majority of plants is able to transform large quantities of inorganic matter into living one. So we know f. i. that the whole body of a poplar is formed of inorganic matter, thus transformed by the wee little bit of living substance, once present in the seed which sprouted to form the poplar.

Where we thus find that all bodies on earth, living as well as non-living ones, are built up by the same elements, where we know furthermore that these very

same primary substances are present on those numberless celestial bodies which a clear night reveals to
us, where we know that changed conditions continuously cause new combinations to be formed between
these elements, where we know that infinitely small
quantities of living matter are able to transform large
quantities of non-living matter to living one, and where
finally we know also that some elements at least, can
directly give rise to other ones, it is not too risky to
conclude that whatever we see — the revelation of every
day — as Prof. BOSSCHA used to express it, is nothing
but the continuous transformation of that one or of those few wonderfull primary substance(s), from which
everything, ourselves included, originated.

How?

As yet we know but little about it.

Physicists, chemists, geologists and biologists, all of
us, lift but the tips of the dense veil which hides the
mysterious workshop of nature, and but all too frequently we have to thank those who come after us
for letting the tip, we lifted, fall down again, to hide
the errors we committed. But not-withstanding all
our stooping, we who try to reveal the underlying
secret of nature, feel, even in the midst of our defeats,
that the fundamental idea on which all our efforts are
based, that the conception of continuity in all what
happens, that the Principle of Evolution, which owes so
much to Charles Darwin, is correct that, as he expressed
it, the „ordinary succession by generation has never
once been broken".

PART II.
THE EVOLUTION OF LIVING BEINGS.

CHAPTER I.

DEFINITIONS OF TERMS USED.

The evolution of living beings is, as follows directly from what has been said in the introduction, but part of the general problem of Evolution, and again the Origin of species is but part of the more general problem of the origin of all those differently constituted Types which people the globe.

Why, we shall see bye and bye.

For the present we will limit our remarks to the origin of species, and more especially to the origin of those species which we now designate as the diploid ones, e. g. which belong to the class which embraces all so called higher beings.

He who ventures to write on the origin of species, ought to define what a species is, so ought he to do who describes species, no matter whether he considers his task finished when the description has been made, or whether he intends to make use of the described species to build up a more or less elaborate system. In other words: the systematist, as well as the evolutionist, ought to state clearly what he means by a species.

As a matter of fact neither of them usually does.

In 1855 Alphonse DE CANDOLLE said already in his Géographie botanique raisonnée (T. II. p. 1068):

Enoncer clairement ses opinions sur la nature de l'es-

pèce est pour un naturaliste l'épreuve la plus redoutable de toutes. Il sait que chaque mot sera pesé, que toute idée nouvelle pourra être taxée d'hérisie et que des notions fausses sur cette base des sciences naturelles, jettent ses travaux de description dans un discrédit mérité.

And DARWIN says in his Origin of Species p. 30:

Nor shall I here discuss the various definitions which have been given of the term species. No one definition has satisfied all naturalists; yet every naturalist knows vague-ely what he means when he speaks of a species.

Unfortunately, it is just the vagueness of this kind of knowledge which has caused endless trouble and there is a good deal of truth in the „badinage" of MORITZI (Réflexions sur l'espèce. Soleure (Solothurn) 1842) who tells of a professor of philosophy of his acquaintance, *qui avait admis l'espèce, comme nous tous, à priori, sans savoir en quoi elle consiste. C'est seulement plus tard,* he continues, *que l'idée lui est venue de se rendre compte de ce sentiment obscur qui l'a guidé depuis son enfance.*

All this fits our case; all theories of evolution have, until quite recently, been guided by a *vague* knowledge of what a species is, and consequently have inevitably been vague themselves.

The remedy is obvious: one has, before one proposes any theory of evolution, to define the terms which one is going to use.

This is not so difficult as may appear at first sight, because we can be guided by the aim which obviously every systematist, no matter of which period or of which nationality, had before him when he tried to establish a species viz, *to bring like to one and unlike to different species*.

If we admit that identity of the different individuals, to be included in one species, is *the* essential quality of a species, and I doubt whether any naturalist will object to this, the problem is quite simple: *to establish a species, one ought to bring together individuals of identical constitution*.

C'est simple comme bonjour; the trouble unfortunately lies in the difficulty to determine identity of constitution.

As a matter of fact, one has long overlooked this difficulty and thought that one could distinguish different constitutions at sight, even without taking very much pain.

This led the older botanists to the acceptance of groups of supposedly identical individuals as species which later generations have shown to consist of a mixture of individuals of sometimes very different constitutions.

One usually expresses this shortly, by saying that the older botanists considered genera to be species.

It is useless to go into the details of these mistakes and their causes; the obvious primary cause is a lack of discriminating power which prevented them from seeing the differences between the individuals, they thought to be indentical.

Such a lack of discriminating power can however not be reproached to LINNAEUS, so that, if identity of constitution can be established at sight, there is every reason to believe, a priori, that the Linnean species will constitute a group of individuals of really identical constitutions. As a matter of fact, one has believed this to be the case for a considerable time.

Linnaeus himself thought so, or at least, doubted it only occasionally and then set his doubts aside; yet he already was convinced himself, that the checking of his conclusions, based on visible differences, necessitated experiments.

Many before him had felt this: JOHN RAY had said already in the 17th Century, that there was no better criterium to distinguish species, than the fact that true species faithfully reproduce their kind by seeds: nulla certior.... quam distincta propagatio ex semine.

This test was frequently, though not under all the necessary precautions, resorted to by Linnaeus who, by applying it, saw that the reproduction was *not* faithfull, that, on the contrary, the different individuals obtained from seeds, supposed to belong to the same species, showed visible differences. Of some of these differences, he could demonstrate the dependency on external conditions and show that they were not transmitted to the offspring, so that they could be neglected.

Such differences he called *varietates minores*, and in order to call the attention of his students to the relative unimportance of such differences, he characterised them as: varietates minores non curat botanicus. Unfortunately, Linnaeus drew from a few experiments and observations the conclusion, that all *small* differences between individuals should he considered to be such uninheritable differences, and that consequently the *degree* of visible difference was sufficient to make out without experiment, whether it was inheritable or not, in other words that an experienced systematist could distinguish at sight whether two or more individuals had the same constitution or not.

This mistaken idea, the famous „systematischer Blick", unfortunately reigns yet supreme in many herbaria and musea of the present time. It has been exposed to ridicule already in 1855 by JORDAN, who said: *„On s'est efforcé de mettre en vogue* une théorie, qui consiste dans l'admission parmi les végétaux *de types* „*spécifiques tranchés et* dans l'hypothèse de *la varia-* „*bilité de ces mêmes types.*

This hypothesis, as Jordan so justly calls it, the hypothesis of the variability of the species, has caused, in my opinion, all the trouble we have experienced in looking for the causes of the origin of species because it *withdrew from systematics as well as from all theories of evolution every firm footing.*

Very rightly again Jordan has said:

Rejeter le critérium de la permanence héréditaire, c'est tout réduire à de simples hypothèses, à l'arbitraire, à la fantaisie des appréciations individuelles, c'est en un mot *donner pour fondement à la science, le scepticisme : ce qui revient à la détruire.*

And the cause of these individual appreciations, cause of this scepticism which refused to accept the existence of clear-cut species and considered all species to be variable, was in the beginning the total neglect of all experiments (by the older botanists), subsequently the insufficient application of the experimental test

2

(by Linnaeus), and finally again the total neglect of all experimenting.

Here Linnaeus also is at fault, because he *knew* that there existed other than non-transmittable differences between the individuals, united by him to one species; he even designated these f. i. the different forms of cabbage which were cultivated at his time, by giving them another name than *varietates levissimas* viz *varietates* tout court.

That he considered these to be of considerable importance, clearly ressorts from his advice to his students: varietates numerosae plurium specierum attente inspiciantur (Am. Acad. I. 1744 p. 55), but later he has forgotten this distinction which he himself made, *fide* his utterance (Philosophia botanica. 100 1751) *Varietas* est planta mutata a causa accidentali, solo, calore, ventis etc.

It is a great pity that Linnaeus postponed the experimental study of the gardenvarieties (his „*varietates*") until he would find time for them, because that time never was found; the existence of transmittable differences within the Linnean species was forgotten and with this the necessity to experiment was forgotten also.

HAECKEL says consequently very rightly, in his Generelle Morphologie:

„*Indessen war man vollkommen zufrieden, wenn man bei einer untersuchten Anzahl höchst ähnlicher Individuen die Uebereinstimmung in allen wesentlichen Characteren festgestellt hatte.*"

One neglected all experiments and frequently does so yet.

Even as late as 1901, WETTSTEIN says in his Handbuch der Systematischen Botanik p. 13:

„Man wird daher als Art die Gesamtheit der Indivi-„duen bezeichnen können, welche in allen, dem Beo-„bachter *wesentlich* erscheinenden, Merkmalen unter „einander und mit ihren Nachkommen uebereinstim-, „men."

It is true that he adds: „und mit ihren Nachkommen uebereinstimmen," but the systematical praxis takes now as little notice of this as it did in 1886 and is perfectly content to establish species on morphological characters only.

Essential („wesentliche") characters are considered to be specific characters, minor ones: varietal characters, but which characters are essential, which minor ones, nobody says.

Such minor forms are called varieties, formae, races, lusus and tutti quanti and if, which happens rarely, one or another investigator takes the trouble to sow seeds of any of them, and finds to his surprise that these „minor" forms reproduce their kind faithfully, this causes not the slightest change in his opinion as to their varietal or racial rank (these two designations usually being used indiscriminately.).

Yet, Jordan has already in 1855 ridiculed this casus positionis in such a scathing manner, that it is hard to understand how this view could be held so long. He said:

Selon les partisans de cette théorie (of the variability of species) les vrais types spécifiques doivent pouvoir être reconnus et distingués entre eux sans aucune difficulté, même sans étude ni effort de la part de celui qui les observe; toutes les formes végétales qui ne se distinguent pas si facilement.... ne sont que des variétés et ne doivent pas être élévées au rang d'espèces. S'il est prouvé qu'elles sont constantes, qu'elles se reproduisent invariablement par le semis de leurs grains, c'est indubitablement que *le type spécifique a alteré en elles par les circonstances locales*, par l'influence des stations ou par tout autre cause.

Ne sait-on pas, disent-ils, que les espèces végétales sont étonnement sujettes à varier; et n'en voit-on pas dans les cultures un grand nombre, qui varient à point de devenir presque méconnaissables? Parmi les variétés des cultures, n'y en a-t-il pas qui sont constantes, que l'on reproduit de leurs graines telles que celles des Blés par exemple et qui *constituent ainsi de vraies races permanentes? Et une excellente preuve* qu'elles sont effectivement telles qu'on les suppose, *que ce sont bien des races, c'est que c'est là une opinion generalement admise.*

With opinions however, Jordan fortunately was not satisfied.

He went at the question in the only logical way, saying: I see differences within the Linnean species; what are these differences? Speculating about this is no use, calling them names, in casu calling them variations, gives no solution, to conclude from multiplicity to the existence of variability is entirely unwarranted; the only right thing to do, is to test their nature by experiment. So he began to examine the flora of France carefully and was soon able to show that differences were observable within *all* Linnean species, that whatever Linnean species is carefully observed in nature, it can be dismembered into groups of individuals with common characters, differing from other groups of individuals with different characters, but also common to these latter ones.

The only logical conclusion so far was: the Linnean species is not a unit, but a mixture of individuals of different constitutions.

But this conclusion had to be proved.

How to do this?

The closest examination, Jordan quite seized this point, would be insufficient to test the constitution of an individual, because if such a thing as variability really did exist, there was no reason to suppose that an individual, modified in an unheritable way, could be distinguished from an individual, possessing the same characters innately, and consequently in a transmittable way.

Jordan *consequently discarded morphological comparison as a criterium for specific purity* and, falling back to Ray, (of whom he may or may not have known) *substituted for it : nulla certior . . . . quam distincta propagatio ex semine.*

This was a great advance, a milestone on the road to knowledge. Carefully collected seeds, each lot from a single individual, protected against contamination by crossing, brought, upon sowing, the important fact to light that *within the Linnean species there are indi-*

viduals of different constitutions and from this, experimentally established, fact Jordan drew the well founded conclusion:

The Linnean species is no species.

Yet, up to the present moment, one has continued to speak of it as such, and this has caused endless trouble. Most decidedly the time has come to break with this wrong designation.

I now propose to do so and to replace the name Linnean species by the neutral name: LINNEON and to define:

A Linneon as the total of individuals which resemble one another more than they do any other individuals.

The different types, faithfully reproducing their kind which can be distinguished within the Linneon, Jordan now called, quite reasonably: species. They have subsequently been called: mikrospecies, Jordanian species, subspecies, small species or elementary species indiscriminately.

The question remains: are they really species e. g. have all individuals belonging to a Jordanian species identical constitutions?

The answer depends on what answer must be given to the question:

Is the standing of the test of faithfull reproduction by seed, proof of specific purity? The answer unfortunately is an emphatic: no.

We know that there are f. i. two kinds of white mice, externally indistinctible, reproducing their kind faithfully and yet of different constitutions, as becomes at once apparent upon mating a female of each kind with the same black male. The youngs of the one female will then be uniformly black, while the young ones of the other female will be uniformly grey. *Breeding true to type is consequently by itself no reliable test for specific purity* and consequently there is no certainty that a Jordanian species is really a species; forms from which we know nothing else than that they breed true to type, may consequently not be designated as species but must receive another name; as such I propose JOR-DANONS.

The question remains: what is a species?

The great advance in hybrid analysis in recent years which allowed us to find out the constitutional differences existing between externally indistinctible forms as the white mice, mentioned above, allows us to give a *definition of the species*, viz:

A species consists of the total of individuals of identical constitution unable to form more than one kind of gametes.

All specifically pure individuals are consequently monogametic. The question remains *how to test for specific purity* and the best applicable, although not always reliable, test is probably the one suggested recently by Davis:

If two individuals crossed together give a uniform F_1, these two individuals are specifically pure.

The explanation is quite simple: each of them produces but one kind of gametes; all children therefore must have the same constitution and consequently must be alike under like conditions.

Some possible objections however must be considered:

1stly. In cases where one and the same individual
forms one kind of male gametes and one, but a
different kind of female gametes, the crossed
offspring of two such individuals would in F_1
be uniform also, because each child is the product of the combination of a female gamete
with a male one and all female gametes as well
as all male gametes are alike, allthough the male
ones differ from the female ones.

Such a state of affairs can however easily be
detected by the fact that the reciprocal crosses
will be different, thus indicating that there
must be something amiss with the specific purity of the tested form. We must therefore *make
the Davis-test more severe* by extending it as
follows:

*Specific purity is indicated by the uniformity
and identity of the F_1 generations obtained by
crossing the individuals to be tested,*RECIPROCALLY.

2dly But in the following case even this more severe
test would remain indecisive viz, if there were
preferential mating between the gametes of two
hybrids, crossed together in such a way, that only
one kind of the different gametes produced by
the one hybrid were able to combine with one
kind only of the different gametes produced by
the other. In such a case the F_1 generations,
even the reciprocal ones, would be uniform and
identical and thus simulate specific purity of
the tested individuals.

In dealing with plants however, such cases

would generally be detected by the presence of a number of aborted ovules and aborted pollen-grains.

Forms showing such abortion are however from the very beginning suspect, so that one does well to exclude them from experiments necessitating the use of pure species.

3dly. *To make the Davis-test as reliable as possible, we must furthermore cross the form to be tested with more than one other form as pure as obtainable.*

This will f. i. be absolutely necessary in the case of albinotic forms, because in these cases constitutional differences remain invisible in the F_1 generations, obtained by crossing two albinos of different constitution, f. i. two of the three kinds of white mice described in Bateson's Principles on p. 75.

In such cases crossing with a colored form reveals the difference.

But from the fact that the F_1 generation of crossed albinos does not necessarily reveal the difference in constitution between those albinos, it is safe to conclude that the possibility exists that the F_1 generation, obtained by crossing two colored individuals, may also, in some cases, fail to reveal certain differences in constitution existing between these individuals.

It will therefore be safest to test for specific purity in the following way.:

1stly. cross the form to be tested with as many supposedly pure species as obtainable, reciprocally.

If the F_1 generations so obtained, are all uniform, and if the reciprocal F_1 generations of each cross are identical, there is reason to suppose the tested form to be specifically pure.

2dly. submit as many of the F_1 individuals of each cross to as extensive hybrid-analysis as possible, in order to test their constitutional identity. If they prove to be of identical constitution in all points tested, this fact, combined with the result obtained sub 1, makes it very probable indeed, that the tested individual is specifically pure.

It must be conceded that even this gives no *absolute* certainty that the form tested is specifically pure in all respects, but it is the best we can do.

But even the simpler Davis-test has the great advantage that we can say, that if two forms crossed reciprocally, give a heterogenous progeny, either one or the other is, or both are, specifically *impure*, while, if the progenies are each uniform, but not identical in reciprocal crosses, there is no specific purity either.

Because in these cases there is proof, that at least one of the individuals crossed, forms more than one kind of gametes, is polygametic and consequently a hybrid.

One class of differences within the species remains to be considered: the non-transmittable effect of external circumstances on the individuals composing the species. That this can be considerable, shows f. i. a comparison of specimens of the same species growing at high altitudes in the mountains, with specimens growing in the plains. These we will indicate as *modifications*.

We thus get the following definitions:

LINNEON: *to replace the term species in the Linnean sense, and to designate a group of individuals which resemble one another more than they do any other individuals.*

To establish a Linneon consequently requires careful morphological comparison only.

JORDANON: *to replace the term species in the Jordanian sense, viz: mikrospecies, elementary species etc. and to designate a group of externally alike individuals which all propagate their kind faithfully, under conditions excluding contamination by crossing with individuals belonging to other groups, as far as these external characters are concerned, with the only exception of noninheritable modifications of these characters, caused by the influences of the surroundings in the widest sense, to which these individuals or those composing the progeny may be exposed.*

To establish a Jordanon, morphological comparison alone consequently does not suffise; the transmittability of the characters by which the form was distinguished, must be proved by experimental breeding.

SPECIES: *to designate a group of individuals of identical constitution, unable to form more than one kind of gametes; all monogametic individuals of identical constitution consequently belong to one species.*

To establish a species, neither morphological comparison alone, nor experimental breeding by itself is sufficient, nor are the two combined; hybrid analysis is required in addition.

HYBRIDS: *to designate all individuals able to produce more than one kind of gametes, e. g. gametes of different constitutions, (in some cases all of these or part of these are non-viable). Hybrids consequently are polygametic.*

In many cases the hybrid nature of an individual can be demonstrated by careful breeding alone, e. g. in those cases in which the uncontaminated progeny is visibly heterogeneous; in many other cases f. i. in all cases of albinism, hybrid analysis is required in addition.

MODIFICATION: *to designate the non-transmittable effect of external circumstances.*

CHAPTER II.

DO DIPLOID SPECIES VARY?

Armed with the definitions given at the end of the first chapter, we will consider this momentous question. All the more generally accepted Theories of Evolution are based on so called variability.

We have already seen that the existence of variability, deduced from the behaviour of individuals belonging to a Linneon, considered to be a unit, is doubtfull. Consequently theories like those of Lamarck and Darwin, the basis of which is the Linneon considered as a unit, are untenable, unless it can be shown that pure species are able to vary.

This species undoubtedly are in an uninheritable way. Each species is liable to modification, but such variability is not meant here, the question is whether a species is variable in a transmittable way or not.

The only sense in which the term „variable" can possibly be applied to a homozygous individual, as all individuals belonging to a pure species necessarily are, is *the demonstration that a homozygous individual can by itself* e. g. without having been crossed, *become heterozygous,* so that an individual, producing until to day only one kind of gametes, would proceed to produce different kinds of gametes to morrow.

In other words: *variability implies the possibility of a monogametic individual becoming polygametic under conditions when all possibility of a cross is excluded.*

This is not inconceivable; it could be brought about in two ways: certain gametes might loose a property, they previously possessed in common with the other gametes, or some gametes might gain a property which they, as well as the other gametes, previously lacked.

Such a processs is *conceivable*, but it would add something fundamentally new to our stock of knowledge; something entirely out of the pale of Mendelism f. i. and to which therefore Mendelian behaviour could give no direct support, although the contrary is believed so frequently, because Mendelism deals with the demonstrated fact that a heterozygous individual produces more than one, a homoygous individual but one kind of gametes, and has nothing to do with the way in which heterozygotism arises, except if it does so by crossing.

But the variability required by evolutionary theories, based on a process of variation, should be able to change a homozygous organism, *without* crossing, into a heterozygous one.

We repeat: this is not inconceivable, but very clear proof for its real existence is wanted, before we are justified to accept it, as sufficient evidence for such an unexpected and novel occurrence.

Now how can we possibly get this proof?

Mendelism can 't give us the clue; *it does not deal with the internal structure of the gametes*, except in so far as *deductions* are concerned, derived from the *fact* that the hybrid arisen from the product of the fusion of two gametes of different constitutions, is able to form more than one kind of gametes.

Mendelism deals with the behaviour of gametes already formed.

As such however, it can be of service, because, although it is unable to show us how the change necessary to bring about the heterozygous condition, took place, it can show us the result viz: the then heterozygous condition of the previously homozygous organism.

If we now *define a mutation as the change of constitution, undergone by a homozygous individual which became heterozygous without having been crossed*, we can say, that PROOF OF MUTATION WOULD BE AT HAND, IF IT WERE SHOWN THAT A HOMOZYGOUS INDIVIDUAL CAN BECOME HETEROZYGOUS WHEN ALL POSSIBILITY OF A CROSS IS EXCLUDED.

Mendelism could show us, that such mutation had taken place, if we were but sure — it is the old difficulty again — of the homozygous condition, e. g. of the specific purity of the material from which the supposed mutants arose. Does the classic subject for mutation: Oenothera Lamarckiana give us proof for the existence of such mutations?

The answer is an unconditional: no.

In the first place, Oenothera Lamarckiana has never been obtained as yet, in a homozygous condition e. g. in a condition that it threw no „mutants"; every Lamarckiana-individual, so far examined, „mutates", was consequently impure already, so that the bringing to light of these mutants was comparable to the bringing to light of the presence of silver in a lead-ore containing silver.

As little as the latter fact shows, that silver arose by mutation from the *element* lead, as little the former shows that the aberrant forms, obtained from selfed O. Lamarckiana, arose from the *species* O. Lamarckiana.

This is the Lamarckiana-question in a nut-shell.

DE VRIES has shown that O. Lamarckiana is a heterozygote and he has *shown* nothing else; all the rest is mere hypothesis.

This very obvious conclusion is corroborated on all sides.

It is easy to prove the heterozygous condition of O. Lamarckiana by crossing it with its „mutants", and by examining the F_1 generation, so obtained, according to the Davis-test for specific purity, as to its uniformity.

By so doing we cross — according to de Vries' conception — two pure species (O. Lamarckiana as well as its mutants being considered to be species by him), and consequently should obtain a uniform F_1 generation.

Now the result proves by no means this contention.

In most cases of such crossing we obtain, what de Vries has called segregation in F_1, which means nothing else than that the F_1 generation is *not uniform,* and from which fact no other conclusion is warranted than that either O. Lamarckiana or the mutant is able to form more than one kind of gametes, consequently is a heterozygote; and as we know that Lamarckiana, as well as the mutants, is able to throw „mutants" it is most probable that *both* are heterozygotes.

This makes it highly probable that O. Lamarckiana only *simulates* a species, but in reality is nothing but a

Linneon e. g. a group of morphologically similar, but constitutionally, different individuals.

That this is really the case has been shown admirably by HERIBERT NILSSON who proved among other things, that individuals of the Linneon Oenethera Lamarckiana produce, after selffertilisation, different „mutants" and in different proportions, from which he draws the only warranted conclusion that different individuals of O. Lamarckiana have different constitutions.

The fact is not challenged by de Vries, but he takes — indirectly — exception as to the conclusion.

So he says on p. 340 of his Gruppenweise Artbildung:

„Die Mutabilität einer reinen Rasse ist keineswegs stets dieselbe; die Ernte der einen Mutter ist oft viel reicher an Mutanten als diejenige einer Nachbarin".

Or in english:

„The mutability of a pure Race is by no means al-„ways the same: the crop of one mother is frequently „much richer in mutants than the one of a neighbour."

We naturally expect to find conclusive evidence for the legitimacy of this startling assertion in the results of new experiments with a race, the purity of which was beyond all reasonable doubt.

Unfortunately this is not the case; de Vries bases this assertion solely upon a creed, and on nothing but a creed: the *supposed* specific purity of O. Lamarckiana.

It is useless to fight against dogmatic creeds; where they lead to, the sentences, following immediately upon the one quoted above, show:

„Von Oenothera scintillans, welche gewöhnlich aus „ihren Samen van 15—40% scintillans Pflanzen er-

3

„zeugt, habe ich einmal eine Rasse gehabt, welche „deren 66—93%, im mittel 84%, hervorbrachtte".

And this, to support the assertion that the mutability of different individuals of a *pure* race can be different!

On the question: how do you prove that different individuals of a *pure* race can be of different mutability? one thus obtains the answer: well do'nt you see? I had once a *group of races* (O. scintillans) in which individuals belonging to *different* races, [1]) behaved differently!

And this is not a slip of the pen on de Vries part!

In a very recent article (Bot. Gazette Nov. 1915) he says on p. 341:

„In order to obtain species of O. gigas yielding a high „percentage of dwarfs from their seeds, I sowed in 1911 „seeds of my *pure strain* [2])", cultivated as biennials, and „fertilised them 1912 by their own pollen in bags".

The result was an offspring, consisting of plants which threw no dwarfs, others which threw 0,6% of dwarfs, others which threw 1.5% of dwarfs, others again which threw 2.3% of dwarfs, yes even plants which threw 17.8% of dwarfs! Yet de Vries refuses to entertain the idea that his O. gigas might have been impure!

Nor causes the fact [3]), that the cross O. Lamarckiana × O. nanella gives a heterogenous F_1 generation, con-

---

[1]) Von O. scintillans, welche *gewöhnlich* von 15—40% scintillans erzeugt, habe ich früher einmal eine *Rasse* gehabt, welche deren.... im mittel 84% hervorbrachtte.

[2]) The italics are mine.

[3]) Hugo de Vries: Ueber amphikline Bastarde. Berichte Deutsch botanische Gesellschaft November 1915.

sisting of Lamarckianas and nanellas in different numerical proportions, depending on the „individuelle Kraft
der Samenträger" (euphemism for: different genotypical
constitution), de Vries any uneasiness as to the justifia-
bility of considering his Lamarckiana as pure; far from
doubting the purity of his strain, he concludes unblus-
hingly from some rather rough experiments, that „die
Kreuzung van O. Lamarckiana mit O. Lam. mut. nanella (sic) liefert, je nach den Kulturbedingungen,
o—90% Zwerge!

The question of the specific purity of O. Lamarckiana, the basis of de Vries' assertions, is not, as it ought to
be to him, an object worthy of the most scrupulous
investigation, but a dogmatic creed.

This involves of course the *belief* in the existence of
mutations thrown by O. Lamarckiana; we can not deny the possiblity that mutations may exist, but as
scientists, we cannot be satisfied with a mere belief; we
want *proof*, especially where it concerns such an entirely novel thing, as the real existence of the, *so far only
surmised, existence of mutations* would be.

The behaviour of O. Lamarckiana gives us not the
slightest cause to suppose that the aberrant forms,
thrown by it, owe their origin to a process of mutation;
they can perfectly well be explained, without ressor-
ting to an „explanation", lying outside of the pale of
experience, by the *simple fact* which the experiments
which have been published all tend to show, viz. *that O.
Lamarckiana is a mixture of heterozygotes of different
constitutions throwing rogues* (the pretended mutants)
by a process of mendelian segregation, in proportions,

deviating from the normal mendelian numbers, by the failure of certain kinds of gametes to produce viable offspring with other kinds of gametes, while certain other kinds, mating together, do produce viable offspring; in other words the normal mendelian numbers are disturbed by preferential mating, a process known to occur in other cases (gametic coupling and gametic repulsion).

All this is furthermore complicated by the death of some kinds of gametes before they are in a position to mate, as is shown by the sometimes very large proportions of abortive pollengrains and ovules, and complicated once more by the death, as shown by RENNER, of certain kinds of embryo's at a very early stage in their development, all or part of this, perhaps either caused by, or complicated through, irregularities in the meiotic divisions of these heterozygotes.

No unprejudiced investigator will accept the behaviour of such an evidently impure form as evidence for the existence of mutation.

The question remains: is there sufficient evidence to accept the existence of such a novel process as mutation would be, in other cases?

The difficulty is again to obtain material of unimpeachable specific purity to experiment with.

We have seen that we possess *no* certain means to prove specific purity in any case, but we know that it is far easier to obtain the recessives in a pure state than the dominants.

Now it is a fact, well worthy of our careful consideration, that whenever so called mutants are described, these are invariably thrown by dominants and not by recessives, which makes it very probable that the dominants throwing them were impure, *because the fact that, „dominants" throw „mutants" and recessives do not, can also be expressed in this way*: that, WHENEVER THE PURITY OF A FORM IS VISIBLE (as in recessives) IN REGARD TO CERTAIN CHARACTERS, IT THROWS NO „MUTANTS", INVOLVING THESE CHARACTERS, WHILE A FORM, FROM WHICH IT CAN NOT BE DETERMINED AT SIGHT, WHETHER IT IS PURE OR NOT (as in the case of a mixture of pure dominants and impure dominants) MAY THROW „MUTANTS".

The only legitimate conclusion which we can draw from the regrettable fact, that it is so difficult to make out with certainty that a certain form is specifically pure in all respects, is that we must require the severest possible test from him who wants to prove the existence of mutations. Knowing how difficult it is, to show that a given form is free from recessives, we must disqualify, a priori, all claims of having proved the existence of mutations, based on the demonstration that a certain form has thrown recessives, no matter in how feeble proportions. In the second place we must refuse to accept as evidence all cases in which the reciprocal F_1 generations, obtained by crossing the supposed mutant with the form from which it arose, are either not-uniform or dissimilar.

In the third place we must refuse to accept as evidence all cases in which the numerical proportions in the uncontaminated F_1 generation of such a cross deviate from the normal mendelian ones, or can only be explained by the acceptance of numerous so-called factors, because all this indicates complication and therefore the possibility of errors of interpretation.

We are therefore, for the present, forced to require from him who wants to furnish proof for such an unexpected novel fact, as mutation would be, *at the very least*:

1stly. that he has previously investigated, to the best of his ability, by hybrid analysis, the purity of the form from which the supposed mutant arose.

2dly. that the supposed mutant, crossed with the species from which it arose, does not behave as a recessive, but either dominates in the homogeneous and identical, reciprocal F_1 generations, or shows in such F_1s, characters intermediate between those of the original species and the mutant; furthermore that the mutant and the form from which it arose reappear in the uncontaminated F_2. generation, either in the proportion 3 : 1 or in the proportion 1 mutant : 2 intermediates : 1 original form.

This must be at present the test required because, so only, it can be proved that a homozygous individual can become heterozygous by itself, in other words can mutate.

As far I as am aware, no pretended case of mutution can stand this test.

There is consequently not the slightest proof for the existence of mutation and DAVIS with whom I agree in the main, goes in my opinion decidedly too far when he says (Science XLII Nov. 5. 1915):

„One may be a mendelian, firmly believing in the „principle of segregation following an F_1 generation, „which is the principal test of mendelism, and still „admit the probability of modifications from time to „time of the stereochemistry of germ-plasma even in so „called „pure lines".

We may admit the *possibility* (because to deny a possibility is unscientific) of such stereochemic modifications, alias mutations, but there is no reason at all to admit the *probability*; all we can do, is to say that the existence of such an inheritable changability of the germplasma would be exceedingly interesting, but that, so long as such proof is not forthcoming, we can take no account of such a mere possibility in any effort to explain evolution.

Evolution has suffered quite sufficiently from the „possibilities" with which it is charged, so that I fear that the addition of another mere possibility might be like the straw that broke the camel's back, e. g. would throw the theory of evolution in universal discredit.

I hear it objected that the appearance of *constant* new forms in small numbers from a certain form is good evidence for mutation, because, if the new form were a hybrid, it would have to segregate. This is of course not true, if the original form is a heterozygote; because then the new form may be the product of the mating of two identical gametes, produced by this heterozygote, and consequently be homozygotic itself.

Mutation can therefore, for the present, be discarded as a factor in evolution; what other kind of transmittable variation has been claimed to exist, by evolutionists?

We all know that a great many persons, up to the present day, express their belief in the inheritance of so called acquired characters e. g. in the occurence of changes in the characters of the individual, caused by external circumstances, and in the transmittal of such changes to the offspring.

Very few among these, yet believe, in the orthodox way, that external influences first change the soma and through this subsequently the constitution of the gametes.

At present the adherents of theories of a transmission of acquired characters, usually mean a direct influence on the germ-plasma, causing a constitutional change in the gametes, which as a matter of fact would be nothing than a mutation, just as mutation is inconcievable without some kind of inheritance of acquired characters.

To this supposition of course applies exactly what we have already said about mutations. The *possibility* can not be denied, but proof is lacking.

In as much as nothing happens without cause, *all* suppositions of transmittable variability, no matter whether we call the effect mutation or by some other name, *have* to assume that, at the end, such variability is caused by external circumstances, *have* consequently to admit the transmittability of acquired characters.

Darwin clearly perceived this, as results from his letter to Semper, written but little more than half a year before his death, (Life and Letters III p. 345) in which he says:

„I still *must* believe that changed conditions give the impulse to variability but that they act in *most cases*, in a very indirect manner. But as I said, it is a most „perplexing subject."

For such an inheritance of acquired characters there is however no proof whatever, and so we must conclude that the existence of transmittable variability has never been proved.

The perplexity of the subject, to which Darwin refers, is caused, in my opinion, by the simple fact that inheritable variability does not exist.

———

CHAPTER III.

SPECIES, LINNEONS, GENERA AND OTHER „HIGHER GROUPS" AND EVOLUTION.

The problem of *Evolution* is not, primarily, as is almost generally believed, the problem of the origin of species, but *is the problem of the origin of all those* INDIVIDUALS *of different constitution, which people the earth.*

Of some of these individuals we can make clear-cut definable groups, by uniting those monogametic individuals which are constitutionally identical, to groups which we may call species, and explain how such species can arise.

But when we find, as we do, that such species are very rare in nature, we recognize that the problem set to the evolutionist is not limited to the origin of these pure species which are but rarely or never met with in nature, but embraces the problem of that vastly greater number of impure individuals by which the earth is peopled. ,

The problem of the species and of its origin is consequently comparable to that of the pure chemical substance and its origin, the problem of the heterozygotes of different constitutions which we find in nature and of their origin is comparable to the problem of the ores found in nature and their origin. Now to gain a good insight in the first part which we might call the chemical side of the problem, it was highly desirable to make groups of indentical individuals and to name these groups: species, because by so reducing the number of cases with which we have to deal, we simplified matters without impairing the exactness of the investigation in the least.

The question becomes different, when we proceed to unite the impure forms found in nature, to *groups* of SIMILAR, *but* NOT *identical individuals* because, by doing this, we only *seem* to simplify the problem while, in fact, we complicate it and impair seriously the exactness of our investigation.

We impair the exactness of our investgation because we form groups which are undefinable, and in so doing, open the door to different *opinions*, as to which individuals should be received into such groups. That- as everybody knows causes great trouble in fixing the limits of such groups as Linneons, Genera, Families, Classes etc. But *what is far worse*, forgetting the nature of these divisions, we are unconsciously and gradually led to believe — as generally is believed — that such groups, which we call higher groups [1]), have necessarily as real an existence as species have, and that consequently, their origin must be explained as an origin of entities also, while, in fact, they may have no real existence at all, but exist in our imagination only.

Sixty years ago already, Jordan has, in a discussion with de Candolle, laid stress on the basic difference between a species as an entity and a Linneon (which he

---

[1]) It were better to speak of larger groups.

considered rightly as a kind of genus, and speaks of it
as such), as a group. About this he said:

„La notion de l'espèce n'est point celle d'un objet
collectif comme l'entend M. de Candolle.... la forme
„spécifique, qui équivaut à l'être, à la substance, est
„identique chez tous les individus d'une même espèce
„et toujours indépendante du nombre.... ainsi le
„premier homme que Dieu a créé renfermait en lui
„l'humanité entière'' [1]).

„En assimilant, comme il le fait, les genres aux es-
„pèces M. Alph. de Candolle ne prend pas garde qu'il
„assimile les catégories, qui renferment les êtres aux
„êtres eux-mêmes.

„Le genre n'a pas l'être, il n'est connaissable que
„parceque notre intelligence le constitue être de raison.
„Il existe dans notre intelligence, mais en dehors d'elle
„ce n'est pas un être, c'est un non-être qui n'a la vérité
„que par les conceptions de notre esprit.''

„Now there is a good deal of truth in this, and Jordan's conclusions are well worth considering, but it is
not the whole truth.

The argumentation looses too much sight of the
fact that *primarily* nature can make nothing but individuals.

If but two homozygotic individuals existed — each
of a different constitution — we would yet be able to
say that there lived two species on our globe, so that it
is true that species are realities „indépendantes du

---

[1]) This example is ill chosen; „l'humanité entière'' is not a single species, as Jordan would doubtless have been the first to have argued himself, if he had not been hampered, in this respect, by religious scruples.

nombre".But this does not away with the fact that a species, in as much as it consists mostly of a number of individuals, usually itself is a group. It is true that by the identity of the constitution of the individuals which compose it, it remains capable of being considered as an entity, but this way of looking at a species, consisting of more than one individual, is dangerous because it leads,almost unconsciously, to the unwarranted conclusion that all the individuals of one species must necessarily be the offspring of a single initial pair of identical constitution, in one word, that species must necessarily be monophyletic. Now this is unwarranted, because the constitution of an individual depends exclusively on the constitution of the gametes, from whose union the individual sprang, and is independent of the source of these gametes. Therefore, if identical gametes can be obtained from different sources — and we will see ,in later chapters, that they can — species need not necessarily have a monophyletic origin.

Such a polyphyletic species, though it, by the identy of the constitution of the individuals composing it, remains capable of being considered as an entity, yet is at the same time not only a group, but even a group of individuals of different origin.

Consequently the real difference between a species on the one side and a Linneon or a genus on the other side, is not so much that the one *may* be, and the other *must* be a group, but is that a species is either a specially constituted individual, always homozygotic, or a group of such individuals of *identical* constitution, while a Linneon or a genus is a group of individuals of *different* constitutions.

And when we have once grasped this fundamental difference, we can easily demonstrate that Jordan was perfectly right when he said that l'espèce est indé-pendante du nombre, while the genus is not. If we reduce a species to one single individual, by extermi-nating all others, that species still remains in existence, but when we reduce a Linneon or a genus to one spe-cies, that Linneon or that genus disappears, because a mixture of different types is no longer a mixture after its reduction to one type.

While thus Jordan is perfectly right, when he states that a genus is a group made by us, it is not certain yet, that he is right also when he continues to say that the Linneon or the genus n'a la verité que par les conceptions de notre esprit, because it is possible that groupings, such as we make, coincide with similar groupings made by na-ture.

Because nature primarily can make nothing but individuals, it does not follow that it cannot se-condarily group such individuals in various ways; as a matter of fact we know that it does, by the fact of the existence of different plant-societies, different flora's and fauna's in different countries etc.

Now if our Linneons, obtained by grouping together morphologically similar individuals, happened to co-ver the groupings made by nature, such Linneons would be something more than mere conceptions of our mind and therefore be well worth investigating.

IF nature makes groupings more or less coinciding

with our Linneons, it is of course improbable that we should *always* have grasped nature's meaning and so it may — if natural grouping really does exist — be expected that some of our Linneons will be mere conceptions (and, in that case, wrong ones) of our mind, while others may have existence in nature.

How this could come to happen, an example may show.

Suppose the different races of mankind were pure-bred, — as our schoolbooks, describing the ancient Normans as invariably fair-haired, blue eyed, beautifully proportioned gods and goddesses, rather than men and women, try to make us believe, they formerly were — then the races of mankind would be as many pure species.

Now suppose the world *were* thus peopled by specifically pure Caucasians, Mongolians, Indians and Negroes, and we were charged to constitute an army of them, containing the men as well as the women, then we could solve this problem in different ways.

Suppose we arranged them into 4 bataillons, each bataillon consisting of one species only, then we would get a Caucasian bataillon, a Mongolian bataillon, an Indian bataillon and a Negro one. Now suppose a superior officer came along and said we had united too many individuals into one bataillon, and charged us to cut them up into bataillons of the perscribed size, to unite such bataillons to proper regiments, and such regiments to an army, we might get four armies of exactly the same constitution as our bataillons were, which could very properly be designated as the Cau-

casian army, the Mongolian army, the Indian army
and the Negro-Army.

Suppose now, on the contrary, that the bataillons
we originally made, had been of the proper size, then
uniting them to regiments, would already cause a
change of constitution, because we would get, say a
Caucasian-Mongolian and an Indian-Negro regiment,
in stead of specifically pure regiments, so that the
army, consisting of such regiments, would have to be
designated as the Caucasian-Mongolian-Indian-Negro-
Army. We could of course also constitute a Caucasian-
Mongolian-Indian-Negro army by dividing, from the
beginning, all four species equally over each bataillon.

Now if we compare these different kinds of armies,
we find that each of the armies of the first kind consists of one species only, that in the army of the second kind the bataillons constitute species, but the
regiments Linneons, and that in the army of the
third kind the bataillons already are Linneons, containing different species, be it, each species consisting
of but a small number of individuals.

Now what does Nature say to such grouping?

Suppose we try to find out. To do this, we discharge.
the men and women, marking each of them by a ta-
too-mark indicating the bataillon to which he or she
belongs, and charge them to mark their progeny to
come, with the same tatoo-marks they got themselves,
each child thus obtaining two tatoo-marks, one from
his father and one from his mother.

Now suppose twenty-five years later, we enlist these
children and arrange them in our armies according to

their tatoo-marks. If the different species have not intercrossed, we will find that, although the armies we now compose, consist of different individuals than 25 years ago, they yet have the same constitution as formerly, although it may not be possible to place all the children in the regiments of their parents, because their tatoo-marks may show that a woman from one regiment has married a man from another regiment, be it of the same species as she-herself. This however does not affect the species, so that nature in this case has kept our species pure, our species thus coinciding with those of nature.

Suppose on the other hand that crossing has occured, but has been limited to the individuals belonging to the same regiments of the second army. Then the tatoo-marks will show this to us, and so we will be able to put the children — though many of them will be crossbred, others of course may be pure because their parents belonged to the same species — into the same regiments to which their parents belonged.

Nature therefore has made the same groupings as we did, her regiments coincide with ours, her Linneons with ours, although each of them contains hybrids which ours did not.

But suppose crossing had been promiscuous beween all the species in the second army, then we would be unable to assign to the new recruits places in the same regiments in which their parents had served, because many of them would carry tatoo-marks of different regiments.

In that case therefore, nature had not made groupings coinciding with our regiments or Linneons; our regiments or Linneons therefore turned out to be purely artificial, to be but conceptions of our mind.

Now to this view may be objected, that in the former case, our Linneons were internally also different from Nature's Linneons, because ours contained no hybrids, while those of nature did. This is true, but we will perceive this the less, the more we consider — as we use to do — certain characters as essential to establish Linneons, certain others as unessential or, as we usually express it, as to be of varietal value only.

If we admit f. i. that the Caucasian-mongolian regiment or Linneon may „vary" in skin-color from white to yellow, and the Indian-negro one from red to black, we will accept the regiments or Linneons, nature sends back to us, to be sufficiently similar to those, we had made ourselves 25 years ago, to consider them to be unchanged except by slight variations, too insignificant to change our opinion.

This is no joke; *it is a fact which we must constantly keep in mind, that we get the Linnean species, „das Art-bild" (our Linneon) only, by choosing certain characters which we call essential as criteria and by neglecting others; if we considered* ALL *characters to be of equal importance we would never have grouped the different types, each Linneon contains, together.*

Of course there are Linneons, consisting apparently of one type only; these arise, as we will see later on, through intra-linneontic crossing and selection of one type, usually the dominant one; crossing without selection leads, as it does in human Linneons, to Linneons consisting of nothing but hybrids and this may lead, as in the case of the Linneon *Buteo Buteo*, to a most bewildering number of types within the Linneon, as splendid material present in the Natural History Museum of Leiden, shows.

It results from what has been said, that nature primarily makes but individuals, but secondarily groups of individuals, and that we will have to investigate whether such groupings, made by nature, are covered by our Linneons or not.

———

CHAPTER IV

THE ORIGIN OF DIPLOID SPECIES.

The problem of the species and of its origin is, as was said in the preceding chapter, comparable to that of the pure chemical substance and to its origin.

Just as we do'nt study the origin of pure chemical substances in nature, but investigate this origin in the laboratory, so the question of the origin of species cannot be tackled in the field, but must be studied in the experiment garden.

It is a curious fact that one has pretty generally thought, that the problem of heredity lies at bottom of the question of the origin of species, while heredity of course can only be concerned in the PERPETUATION of a species once formed, because heredity implies the unchanged transmittal of the properties of the parents to the offspring.

That this confusion has arisen, is due to the fact that most organisms are not specifically pure, and consequently get children more or less *similar* to them, but not *identical* with them.

This similarity has gradually become looked upon as a sufficient criterium for heredity, just as the similarity of the individuals composing a Linneon, has been looked upon as to be sufficiently close to allow one to consider a Linneon as a species.

In this way, the conception heredity has become as

inexact as the conception species used to be, and such inexactness is unconsciously carried forward even when, in itself exact, methods gradually evolve.

So it has even tainted Mendelism, for strictly speaking, mendelian behaviour has nothing to do with heredity and to speak of Mendelian heredity is, au fond — I beg Mr. Bateson's pardon — nonsense.

If an organism shows mendelian segregation, it shows no heredity of the character-complex it possesses, but just the contrary: disintegration of this complex and distribution of the vestiges of it over different individuals e. g. non-transmittability of this complex.

The problem of the origin of species is not the problem how a species is perpetuated — how its characters are inherited by its offspring — but is the problem how a species (or a pair of them) can give rise to a species (or to more species) differing from it (or from them). This problem has been solved by Mendel who showed, that by mating gametes of different constitution — brought together by crossing different species — zygotes are formed from which individuals arise, able to form a number of gametes of different constitution, from whose matings new species can arise.

By crossing two monogametic individuals of different constitution, one consequently obtains a polygametic hybrid which is the source — and up to the present *the only known source* — of the origin of new forms, some of which are heterozygotes, others of which are homozygotes e. g. new species.

By isolation of such homozygotes in the experiment

garden, and by selfing them or by mating them with other individuals of identical constitution, but of different sex, we can multiply them and thus obtain new species consisting of as many individuals as we choose to raise.

Evidently it does not matter from which species the gametes which form the zygote, that sprouts to the polygametic hybrid, are derived; it is not even necessary that these gametes should be derived from pure species; they can just as well be derived from hybrids, because the result has nothing to do with the origin of these gametes, but only with their constitution.

Consequently new species can originate, as well in a monophyletic as in a polyphyletic way.

The spot, where the mating of such gametes takes place, is of course as indifferent to the effect as the origin of these gametes, consequently species can arise polytopically e. g. the same species may be born at different spots. That this is a fact can easily be shown.

By crossing (Baur 2d edition p 94) a homozygous yellow *Antirrhinum majus* with a homozygous red *A. majus* we obtain among the segregates in F_2 homozygous yellow-red species, which however we can obtain just as well, by crossing a heterozygous red form of the Linneon *A. majus* with a heterozygous pale yellowred form of the same Linneon.

This as an example, that the same species can arise in different ways.

That the same species can arise at different spots and at different times, is ofcourse an experience of the commonest sort. A rather striking example is furnished by the *Petunias* with green-rimmed petals which were obtained about 1830 in England, and again in 1914 in my garden at Bennebroek, by crossing the same Linneons as were crossed 80 years ago in England, viz *Petunia violacea* and *Petunia nyctaginiflora*. It is of course equally indifferent to the question of the origin of species, whether there exist other kinds of segregations than Mendelian ones or not; all that is necessary, is that new homozygous combinations arise finally from a cross. This can evidently even happen without segregation, if two heterozygotes, containing among their different gametes some identical ones, mate, and two of these identical gametes form together a zygote.

New species consequently are the result of the mating of identical gametes, derived, usually indirectly, from the mating of two gametes of different constitution (by crossing heterozygotes or homozygotes of different constitution) *or derived directly, from the cross of two heterozygotes, having among their otherwise different gametes some identical ones in common.*

CHAPTER V.

THE PERPETUATION OF THE NEW SPECIES.

As every new individual is the result of the mating of two gametes, forming a zygote, and as a species originates either from one, or from a few of such zygotes, multiplication is necessary to perpetuate the species.

The transmittal of the constitution of the parent to the offspring we call heredity.

What do we know of it?

Unfortunately: absolutely nothing.

Heredity of course, deals in the last instance with the constitution of the gametes, and of the constitution of these gametes we know nothing.

One has tried to get at the constitution of the gametes by mating gametes of different constitution, and by recording the behaviour of the zygote so obtained.

The results show that gametes derived from individuals, differing from one another in one respect only, f. i. from plants, differing but slightly in the color of their flowers, give other results, than follon upon the mating of gametes derived from individuals which differ in several respects.

If we mate two gametes, obtained from parents who differ in one respect only, the hybrid produced is digametic e. g. forms only two kinds of gametes, one kind of which is identical in constitution with the gametes produced by one of the monogametic parents of the hybrid, the other kind being identical in constitution with the gametes produced by the other parent.

This of course teaches us nothing, as to the constitution of the gametes; it looks as if the two mated gametes remain side by side in the zygote, and form, linked hand in hand so to speak, the hybrid, to say good bye to one another, as soon as the vegetative development of the hybrid comes at an end and it proceeds to form gametes itself.

Cytology however teaches us that there is not such a loose linkage between the gametes in the zygote, as we would have to accept on the ground of this explanation, but that, on the contrary, a very intimate fusion of the bodies and even of the nuclei of these gametes takes place, in which their individuality, so far as we can see, gets lost with the exception only of the individuality of the chromosomes.

As the gametes evidently regain their individuality as the hybrid proceeds to form gametes, these latter being identical to those, united in the zygote, from which the hybrid arose, it looks as if the supposition of a loose linkage of the gametes in the zygote were after all correct and that, even during fusion, their individuality could be upheld unimpaired by their chromosomes, retaining their individuality.

But there is some uncertainty — and this must not be overlooked, in the assignement of such exclusive importance to the role of the chromosomes, as this interpretation implies — because it is founded on the mere supposition that the essence of the individuality of a gamete lies in its chromosomes, while the only reason we have for this supposition is, that we can make these chromosomes visible and observe their behaviour during nuclear fusion and mitosis.

It must however not be forgotten, that there may be f. i. in the cytoplasm of the fusing gametes particles wich retain their individuality, just as well as the chromosomes do, butwhose behaviour remains intracable to us, because we lack the means to make them visible.

If such particles did exist, it would of course be inadmissable to ascribe to the chromosomes the exclusive ability to uphold the individuality of the gametes.

Yet — taken, all in all, we know of chromosomes especially the fact, that where gametes differ greatly in size, the quantity of cytoplasm introduced by the male gamete, is insignificantly small compared to that of the ovum, and yet reciprocal crosses frequently give the same result — there is certainly good reason to consider the chromosomes to be very important constituents of the gametes, so that there is nothing against the supposition that they are able to maintain the individuality of the gametes, even during their fusion in the zygote.

And with this retention of the individuality of the gametes through their chromosomes, the production by the hybrid of a number of gametes, half of which is identical with the gametes of the one, half of which is identical with the gametes of the other parent, is in full accord.

As soon however, as we proceed to mate gametes derived from individuals, differing from one another in more than one respect, this hypothesis of an individuality of the gametes, maintained unimpaired during the process of fusion, can no longer be upheld, because then the hybrid, sprouting from the zygote, no longer forms but two kinds of gametes, but several kinds, all of which, except two, consequently must necessarily be different from the kinds of gametes, produced by the parents.

How did this change come about?

We know nothing about it.

All so called explanations, all hypotheses of separate bearers of the several characters of the individual, all surmises of so called hereditary units which retain their individuality during nuclear fusion, about in the same way as the chromosomes do, and which, after the opinion of some, form parts of the chromosomes, and, by their interchangability, make new combinations, and consequently formation of a large number of different gametes possible, have their footing in Darwin's hypothesis of pangenesis, no matter whether we call such units pangens, gens, factors or by any other name.

Their existence is a mere supposition; the so called fact of the fragmentation or pulverisation of the chromosomes during synapsis, held by some to give considerable support to their real existence, is in itself far from having been satisfactorily established.

After all, the chief reason for the making of the hypothesis of pangens, is the creed that the smallest particles of living matter must necessarily be living themselves, because life is something so fundamentally different from non-living, that living substance must necessarily be fundamentally different from non-living substance, in its whole composition also.

This however is a mere creed; it is perfectly conceivable, that life is the resultant of forces in very complicated chemical bodies, mutatis mutandis in the same way, as heat can be the resultant of forces, in a mixture of less complicated chemical bodies.

As little as in the latter case each particle of the chemical bodies which we mix, need be hot, as little each particle in the former cases, need be living. The fact demonstrated by BECQUEREL of Paris, in KAMERLINGH ONNES Laboratory at Leiden, that seeds, enclosed in tubes devoid of air, immersed during 3 weeks in fluid air, and after that during 77 hours in liquid hydrogren, thus having been exposed to a cold of 190° till 253° Celsius below the freezing point of water, sprouted all the same, after having subsequently been kept for a year and a half in a vacuum, goes far to throw grave doubt on the living nature of the protoplasm, present in those seeds.

Deprived of water and gases, and under a pressure of almost nihil, exposed to extreme cold, this plasma must have lost its colloidal phase, and can not have shown even the slightest traces of physical or chemical signs of life; we can not designate such a condition, scientifically speaking, by any other name than death.

Yet this plasma is able to show afterwards all normal appearances of growth etc., so that there is every reason to believe that a system of in itself non-living particles can, under favorable conditions, interact in such a way that life results.

The conception of entirely independant hereditary units, retaining their individuality unter all circumstances which by temporary linkage, like shuffled sticky, differently colored, lumps of sugar, give rise to new patterns, comparable to new species, reveals only the naivité of those minds, who consider this a satisfactory explanation of the very complicated interactions which must take place in such a wonderful substance as two fusing germ-plasma's

What happens in the zygote is doubtless a very complicated chemical, not a mere physical, process, as the shuffling of different independant, hereditary units would be.

Yet, even so rough a scheme can be of service, and has been of service, to assist us in gaining a first general insight in what may happen, but when we forget the initial inadequacy of this scheme, and especially when we proceed to hide the inadequacey of the explanation, by all kinds of suppositions, notwithstanding the results of the crosses so ,,explained'' ought to have warned us against believing in the adequacy of this explanation, we get on very unsafe ground. So f. i. in those cases, in which we are trying to explain the shortness of hair, now by the loss of a factor, then again by the addition of a factor, inhibiting the effect peculiar to a length factor supposed to be present, we run great risk to consider our duck a goose, and to gain unwarranted confidence in a scheme which — at the most — can be, but even need not be, correct in a very general way.

And this scheme has already obtained an exagerated value in the opinion of many mendelians who forget, that it is, at the most, a very rough diagrammation of the complicated happenings during the interaction of two mixed germplasmas of different constitution, so that one has even carried it forward, to interpret what happens, when two germplasmas of *identical* constitution fuse and concluded that *every* germplasm is a mixture of independant hereditary units.

Now of this we know absolutely nothing.

The mistaken idea that organisms, showing mendelian behaviour, warrant some such kind of a conclusion, arose from the fact that nearly all organisms, met with in nature as well as under cultivation, man included, are hybrids which were mistakenly considered to be specifically pure, so that their behaviour was unconsciously held to be that of specifically pure organisms, while it was that of hybrids; so it happened that segregation was mistaken for heredity.

Yet, if one wanted to make out the way in which a pure human species, say one with fair hair and blue eyes, transmits the fairness of its hair and the blueness of its eyes to its progeny, one would get no aid whatever from investigation along mendelian lines; while if one investigates the transmittal of eye-color or hair-color in such a society of hybrids, as we are, one finds the laws by which these hybrids distribute their differently constituted gametes over their offspring and is apt, if one forgets the hybrid nature of one's material, to mistake this distribution for heredity, and to speak of Mendelian segregation as of mendelian heredity.

Whether the charactercomplex of a homozygous individual is disintegrated, at the moment of the formation of the gametes, and subsequently rebuilt when two of these gametes fuse, so as to form a zygote, or whether it is never desintegrated at all, and consequently *really* inherited, we know not.

This of course is of importance in connection with the possibility of the inheritance of engrams in the sense of SEMON, because such a thing would be at least conceivable if the charactercomplex were inherited as an entity, but becomes almost inconceivable if, during the process of formation of the gametes and during their subsequent fusion, the engram would first have to be disintegrated and later have to be rebuilt.

We know, unfortunately, absolutely nothing of the way in which a homozygous organism transmits its characters to its offspring, and consequently we know absolutely nothing of heredity.

If we define heredity as the transmittal of the charactercomplex of a homozygous individual as an entity to its offspring, we do not even know whether heredity exists or not.

We are thus not even in a position to say, whether a homozygous individual is an entity or a complex of more or less independant units; it might very well, for all we know, be the necessary product of a germplasma consisting of a definite chemical substance f. i. of a definite albuminous body, in a similar way, as so many cristals are the necessary shape-products of definite pure chemical substances.

The specific shape might therefore be the expression

of a specific chemical substance ;in suppport of this, one might bring forward the established fact that different „species" of fishes are characterised by specific protamines in their male gametes.

I wish it to be distinctly understood that I am neither pleading for the view that a homozygous individual is an entity, nor for the view that it is a complex of more or less independant units; I am simply stating the fact, regrettable as it is, that we know nothing about it.

CHAPTER VI.

HOW TO GET AT THE CONSTITUTION OF
THE GAMETES.

It is perfectly correct that the constitution of the gametes is basic to all questions of heredity, and may have to do, not only indirectly as it certainly does, but also directly, with the question of the origin of species.

DE VRIES showed great insight in questions of evolution by maintaining that the final cause of the origin of species lies in the constitution of the gametes, but his mistake (Quis caruit erroribus?) was that he overlooked the fact, that the cause of evolution does not necessarily lie in the behaviour of a single gamete, but might lie — as it does in my opinion in diploid species at least — in the interaction of two gametes of different constitution.

That we complicate matters—probably unnecessarily — by trying to get at the constitution of the gametes, through the study of diploid organisms, is plain; we ought to begin with the study of haploid organisms.

Among these, mosses offer special advantages because the mossplant — being the haploid generation — is the result of a single gamete, while the higher plants — being diploid generations — are the result of the interaction of *two* gametes.

If therefore, we ever succeed in changing the constitation of a gamete experimentally, the result will be

5

easier seen in the case of the mossplant, the result of a single gamete, than in the case of a diploid organism because here the result is apt to be obscured by the necessity of such a gamete to mate with another one, of which we do not know whether it is changed in the same sense, or not.

The advantage of haploid organisms above diploid ones is, in general, that we can choose directly the gametes which we want to mate, by pairing morphologically different haploid generations, or better still — as in the case of *Spirogyra* — by choosing and mating morphologically different gametes, while in the case of diploid organisms we can only pair diploid individuals, and have no means to find out which of the gametes produced by these, finally fuse, except by judging after the result: the character of the diploid organism, sprouting from the zygote thus obtained.

So the study of haploid organisms may very well be destined to give us the clue to the final causes of the origin of species.

For the present we know but very little about them; of some known facts I will come to speak at the end of this sketch.

Experiments with mosses, seem to me, to be highly recommendable.

For the present we will have to stick to known facts, and therefore will continue our discussion of the origin of the differently constituted types of diploid organisms which people the globe.

CHAPTER VII.

HOW CAN GROUPS OF DIFFERENTLY CONSTITUTED TYPES FORM A LINNEON, SIMULATING A SPECIES?

As we saw in the fifth Chapter, Mendelian behaviour and heredity are different things, consequently heredity and evolution may very well be quite different problems.

I would not be at all surprised if definite proof were some day forthcoming that heredity and evolution are even antagonistic forces. Evolution, as far as at present known facts allow us to judge, is caused by the mating of gametes of different constitutions.

The resulting hybrid gives, after segregation, rise to a certain number of homozygous individuals whose constitution is inherited by their offspring, which then constitute a species, perpetuating its kind faithfully, up to the moment that a new cross intervenes.

At the moment heredity sets in, evolution therefore stops, and at the moment evolution sets in heredity stops.

Now in our experiment garden we have it in hand, to let evolution begin and stop at our will: by executing a cross we start evolution, by weeding out the heterozygous combinations, obtained by the mating of the differently constituted gametes in F_1, and by keeping only the homozygotes and isolating each of these, we stop evolution and let heredity take its course.

Now this is different from what happens in nature; there, no homozygous form is ever sure of being à l'abri of a cross, except if *strict* selffertilisation reigns supreme which it but rarely, if ever, does. Consequently we must look into the question what is likely to happen *in nature* after a cross.

Calculation of what is likely to happen if strict selffertilization follows a cross between organisms differing in one or more respects.

If strict selffertilization takes place, the progeny of each homozygous individual necessarily must be homozygous also, so that every homozygous individual starts a species. The heterozygotes on the other hand, will continue to segregate in all subsequent generations, and give rise to a mixture of homozygotes and heterozygotes.

The principles which underlie the matter are therefore, in Jenning's words, the following:

(1) In self-fertilized organisms, all characteristics that become once homozygotic, remain homozygotic for ever after, since there is no method in selffertilization of introducing a gamete that is diverse in this respect.

(2) Characteristics, heterozygotically represented become homozygotic in a certain proportion of the offspring.

The problem becomes essentially this: in what proportion do the heterozygotic characters become homozygotic, and how great a proportion of all the organisms will therefore have become homozygotic after a given number of selffertilizations?

Jennings, American Naturalist 1912 pp. 487 seq. calculates this as follows:

„Suppose that we begin with an organism in which all separable characters are heterozygotically represented.

1) „*Consider first a single pair of such alternative cha-*„*racters, which* we may call $\left\{ \begin{smallmatrix} A \\ a \end{smallmatrix} \right.$. The gametes pro-„duced will be A, a, A and a, and when these com-„bine in all possible ways, they give zygotes AA +„Aa + aA + aa; that is two homozygotes and two „heterozygotes. Thus, the selffertilization of such „an organism gives $^1/_2$ the progeny homozygotic „(with respect to this characteristic) $^1/_2$ hetero-„zygotic"

„If we let x = the proportion of homozygotes, y = the proportion of heterozygotes (with respect to one character) then after the first selffertilization

$$x = {}^1/_2 \text{ of all}$$
$$y = {}^1/_2 \text{ of all}$$

„Now, after the next selffertilization, of course the „homozygotes x remain pure, so that half of all „the progeny are still homozygotes on this account. „The heterozygotes y of course again break up, in the „way already set forth, one half into x, the other „half remaining y. Since y included half of all, this will „give $^1/_2$ of $^1/_2$ (= $^1/_4$ of all) as x, $^1/_2$ of $^1/_2$ (= $^1/_4$ of „all) as y.

„So the total proportion for the homozygotes x „becomes after the second fertilization:

$$x = \tfrac{1}{2} + (\tfrac{1}{2})^2 = {}^3/_4$$

while

$$y = (\tfrac{1}{2})^2 = {}^1/_4$$

„This process is repeated after each fertilization, „so that if there are n fertilizations in succession, „the total number of homozygotes x, becomes „$x = \tfrac{1}{2} + (\tfrac{1}{2})^2 + (\tfrac{1}{2})^3 \ldots$ up to $(\tfrac{1}{2})^n$.

„This expression reduces to $x = \dfrac{2^n - 1}{2^n}$ where n „is the number of fertilizations.

„For the heterozygotes, y, on the other hand the „formula is simply

$$y = (\tfrac{1}{2})^n$$

„These then are the formulae in case we deal with „but one pair of characters. They express:

1) the proportion of all the organisms that will be homozygotic (or heterozygotic as the case may be), after a given number of n fertilizations.

2) also they of course express the relative probability for a given case, as to whether it shall be homozygotic or heterozygotic.

2) „*When we are to deal with two or more pairs of charac-* „*ters*, the problem may be attacked in two ways. „One is by the general principles of probabilities; „the other is by analyzing the case of two or more „characters, in the way exemplified above.

„The two methods give the same results.

„The first method is by far the simpler. It is „merely an application of the principle that when „we know the probability for each of two or more „things separately, the probability that all of them

„shall happen is the product of the separate pro-
„babilities for each.

„Now we know that the probability x for the
„homozygotic condition with respect to one cha-
„racter is:

$$x = \frac{2^n - 1}{2^n}$$

„For two characters it is then:

$$\left(\frac{2^n - 1}{2^n}\right) \times \left(\frac{2^n - 1}{2^n}\right); \text{ or } \left(\frac{2^n - 1}{2^n}\right)^2$$

„For three characters it is of course.

$$\left(\frac{2^n - 1}{2^n}\right)^3,$$

„and in general, for any number m of characters, the
„probability x for pure homozygotes (or the pro
„portional number of pure homozygotes) is

$$x = \left(\frac{(2^n - 1)^m \cdot}{2^n}\right)$$

„By similar reasoning, the proportion of all the or-
„ganisms that will be heterozygotic with respect to
„all the m characters is

$$y = \left(\frac{1}{2}\right)^{nm}$$

„With two or more characters, there will be of course
„a considerable number of the organisms that are
„homozygotic with respect to some characters,
„heterozygotic with respect to others.

„If we call the proportion of these z, then

$$z = 1 - (x + y)$$

„And if we let v be the total proportion that con-
„tains any heterozygotic characters (so that $v =$
„$y + z$), then

$$v = 1 - \left(\frac{2^n - 1}{2^n}\right)^m = \frac{2^{mn} - (2^n - 1)^m}{2^{mn}}$$

„These formulae may readily be deduced algabrai-
„cally, or verified, by a detailed analysis of a case of
„two or more characters. It may be worth while to
„indicate the method followed, by taking up the
„simpler case of two pairs of characters. Call these
„$\{^A_a$ and $\{^B_b$. The gametes formed are AB, Ab, aB
„and ab.

„When these combine in all possible ways (as
„indicated in the diagrams given in Bateson's Mendelism), these give the following results:
1 AB AB + 1 Ab Ab + 1 aB aB + 1 ab ab + 2 AB ab +
    2 Ab aB + 2 AB Ab + 2 AB aB + 2 Ab ab + 2 aB
    ab = 16.

„It will be observed that of the entire 16, the
„first four are pure homozygotes, the second four
„are pure heterozygotes (heterozygotic with res-
„pect to both characters); while the last 8 are mixed
„(homozygotic with respect to one character, hete-
„rozygotic with respect to the other).

„Letting x = pure homozygotes, y = pure hete-
„rozygotes, z = mixed, we find thus that:
$x = \frac{1}{4}$, $y = \frac{1}{4}$, $z = \frac{1}{2}$, of all.
„Now, bij an analysis of the sort already given, it
„will be found that at the next self-fertilization, x
„remains x; y breaks up, $1/4$ of these becoming x,
„$1/2$ of these becoming z, and $1/4$ remaining y; z breaks
„up, $1/2$ of these becoming x, $1/2$ remaining z.

„Now, when we recall that before the second ferti-
„lization x was $1/4$; y $1/4$, and z, $1/2$ of all, we see from

„the above that after the second fertilization:

$$x = \tfrac{1}{4} + (\tfrac{1}{2} \times \tfrac{1}{2}) + (\tfrac{1}{4} \times \tfrac{1}{4}) = \tfrac{9}{16} = \left(\frac{2^n - 1}{2^n}\right)^2$$

$$y = (\tfrac{1}{4} \times \tfrac{1}{4}) = \tfrac{1}{16} = (\tfrac{1}{2})^{2n}$$

$$z = (\tfrac{1}{2} \times \tfrac{1}{2}) + (\tfrac{1}{2} \times \tfrac{1}{4}) = \tfrac{3}{8} = (\tfrac{1}{2})^n + (\tfrac{1}{2})^{n+1}$$

„These are the formulae for x and y that were ob-
„tained by the other method (since here n and m
„are each 2). This method however gives in addi-
„tion a direct formula for z.

„It is easy to verify the formulae for three pairs
„of characters, though of course the conditions be-
„come here somewhat more complex.

„We may now summarize our formulae, and show
„the results they give in certain examples.

„Let $x =$ the proportional number of organisms
„that are pure homozygotes (with respect to all
„the characters considered).

„$y =$ the proportion that are heterozygotic
with respect to all the characters
concerned,

$z =$ the proportion that are mixed,

$v =$ the proportion that have any heterozygotic characters.

„Then, if $n =$ the number of successive self-fertili-
zations and $m =$ the number of pairs
of characters,

$$x = \left(\frac{2^n - 1}{2^n}\right)^m, \tag{1}$$

$$y = (\tfrac{1}{2})^{mn}, \tag{2}$$

$$z = 1 - (x + y), \tag{3}$$

$$v = 1 - x = \frac{2^{mn} - (2^n - 1)^m}{2^{mn}} \tag{4}$$

Examples : (1) Suppose that there have been 8 self-
„fertilizations, and that we are dealing with 10 pairs
„of characters.

„What proportion x of the organisms will be ho-
„mozygotic with respect to all the 10 characters?
„What proportion will be homozygotic with respect
„to any given one character? To any given two or
„three?

„Taking first the case for the entire 10 characters, by
„formula (1)

$$x = \left(\frac{2^8-1}{2^8}\right)^{10} = \left(\frac{255}{256}\right)^{10} = \log. \ 9.9830020 = 0.961617.$$

„Thus, out of 100 individuals, somewhat above 96
„would be pure homozygotes; or by formula (4), but
„one in 26 would be heterozygotic in any respect
„(v = 0.038383).

„With respect to any one character formula (1) gives:

$$x = \left(\frac{2^8-1}{2^8}\right)^1 = \frac{255}{256} = 0.99609375,$$

„so that all but 4 in 1000 would be homozygotes with
„respect to that character"

„In the same way we find that with respect to any
„two characters the proportion of homozygotes
„would be 0.9922; with respect to three, 0.9883; with
„respect to four, 0.9845, etc.

(2) Suppose that there are 20 pairs of characters,
and that there have been 20 selffertilisations. Then:

$$x = \left(\frac{2^{20}-1}{2^{20}}\right)^{20} = \left(\frac{1.048575}{1.048576}\right)^{20} = \log 9.9999957 = 0.999998.$$

„That is, of a million individuals, all but two would be
„pure homozygotes.

„It thus appears that if the number of separable „hereditable characters is not very great (say not „ above 100), while the organism has been selffer- „tilized for many generations, it is to be expected „that practically all of the organisms will be homo- „zygotic with respect to all their characters, they „will be „pure homozygotes"

In other words this calculation leads to a result resembling that of a Linneon, like Triticum vulgare, consisting of a large number of different pure lines.

Now let us see what the calculation of what happens in an intercrossing population of segregates from a cross, will teach us.

Calculation of what is likely to happen, if free intercrossing follows an initial cros between organisms, differing in one or more respects.

This has been calculated repeatedly, always with the same result f. i. by Baur 2d edition p. 123—129. The simplest calculation I am acquainted with is by Reimers in a forthcoming article on „Die Bedeutung des Mendelismus für die landwirtschaftliche Tierzucht" which will soon be published by Martinus Nijhoff in the Hague. He argues about in the following way:

1. The parents differ in one factor only.

If we call this factor A and its absence a, all F_1 individuals must be Aa. In F_2 we will find the same proportions as after selffertilisation, because it is of course indifferent, whether the eggcell A or a is fertilised by Pollen A or a, derived from the same, or from any other plant. It is the constitution of the pollen, not its source which causes the effect.

In the second generation the proportions of the differently constituted plants will therefore be 1 AA : 2 Aa : 1 aa.

If each of these plants forms the same number of gametes, say four, we will get 4 A gametes from the AA plant, 4 A gametes and 4 a gametes from the two Aa plants and 4 a gametes from the aa plant so that we get

$$
\begin{array}{cc}
8\ \text{A} & 8\ \text{a} \\
8\ \text{A} & 8\ \text{a} \\
\hline
\end{array} \times
$$

$$
\begin{array}{cc}
64\ \text{AA} & 64\ \text{Aa} \\
64\ \text{Aa} & 64\ \text{aa}
\end{array}
$$

which means the same proportion as in F_2, viz 1 AA : 2 Aa : 1 aa.

The proportional frequency of the different combinations consequently remains the same, in consecutive Generations, if promiscuous mating takes place.

2) *The parents differ in several factors.*

In this case the proportions of course must remain the same also in subsequent generations, if the factors are independent in respect to one another, because then, the same applies to each set of two factors, as was calculated above for the set Aa.

We may therefore conclude:

Promiscuous mating brings no change in the proportional frequency of the different possible combinations in subsequent generations; there is no loss of heterozygotes in favour to homozygotes, as happens in the case of habitual selffertilisers.

This agrees fully with Baurs results, calculated on the basis of promiscuous mating of white and black mice, about which he says:

„Ueberlassen wir nun eine Population van dieser Zu-„sammensetzung sich zur weiteren panmiktischen Ver-„mehrung, so werden *alle folgenden Generationen* das „gleiche Zahlenverhältniss zwischen den weissen (aa) „und den beiden sorten (AA und Aa) von schwarzen „Tieren aufweisen."

These calculations consequently lead to the result:

Selffertilisation after a cross leads to a considerable proportional increase of the homozygotes in subsequent generations, random mating has no such effect: the proportion of homozygotes to heterozygotes remains the same in successive generations.

Now all such calculations suffer from the impossible supposition that all individuals born, reach maturity, form an equal number of gametes, and that all these gametes mate and form viable zygotes, by which kind of calculation — as is well known — a human couple, in the possession of one cock and six hens, can look forward not only to a prosperous, but even to a wealthy life.

Unfortunately, the ideal conditions on which such calculations are based are never realised, so that the calculated proportion of homozygotes to heterozygotes is valueless, also because strict selffertilisation never persists in nature, even not in such „typical" self-fertilizers as wheat.

So NILSSON-EHLE showed that Pudelweizen can be

inadvertently crossed to the extent of 1 % by other forms, RIMPAU found similar numbers in the case of four-rowed barley, and MAYER-GMELIN even higher ones in beans.

Yet, in a very general way, the difference in behaviour between selffertilizers and individuals, mating at random after a cross, is correct.

Now it is a fact, fully worth our attention, that — in a general way also — what we find in nature, agrees with these calculations, as far as selffertilisers are concerned, but, *as a rule*, does not agree when random-mating occurs.

Selffertilizing Linneons consist really of a great number of distinct forms — though generally not so astonishingly pure as the calculation would make us suppose — while Linneons, within which random-mating occurs, usually have a rather uniform aspect.

Of course there is no sharp limit between these two categories of Linneons, and this cannot be expected either because strict selffertilisation occurs as little as absolutely promiscuous mating, and what seems, at first sight to be uniform, proves sometimes to be multiform on closer examination.

So f. i. R. E. LLOYD (The Growth of Groups in the Animal Kingdom. Longman Green & C°. 1912. 185 pp.) fide American Naturalist 47. 1913, on studying rats in connexion with the plague problem in India, found that small groups of rats, differing in some respects from the forms regarded as typical, occur here and there. But even if we take full account of this lack of a sharp limit between self-fertilisers and randommaters, the difference between the Linneons, consisting of self-fertilisers, and those consisting of freely intercrossing types, is too apparent to be explained away.

Now what is the reason of this difference?

If it is correct that mutation does not exist — and a sentence in Nillson-Ehle's article on: „Gibt es erbliche Weizenrassen mit mehr oder weniger vollständiger Selbstbefruchtung? Ztschr. f. Pflanzenzüchtung 1915 p 5., gives additional support tot his contention:

„In der Nachkommenschaft recessiv-merkmaliger F_2 individuen „von Weizenbastardirungen kommen nicht selten vereinzelte Indi„viduen mit dem dominanten Merkmal vor und ihre Zurückfuh„rung auf Vicinismus (crossing with a neighbour of another form) „kann, angesichts der bei diesem Versuch gefundendenen oben „mitgeteilten, Zahlen keineswegs als unbegründet angesehen wer„den",

then the only way by which the number of homozygotes can increase in succeeding generations of a freely intercrossing community is: selection.

So that, if it is correct, that a Linneon is the rest of what arose from a former cross, the fact that Linneons inside of which free intercrossing occurs, obtain generally a more or less uniform aspect, is definite proof for Darwin's contention that selection, at least intralinneontic one, in nature, is a fact.

It must, of course, be argued that selection will occur as well within self-fertilizing Linneons as within Linneons inside of which free-intercrossing occurs, but this must then lead to a different effect.

Now let us look into this question.

Inside of a community of different types, originated from a cross, which do not further intercross, we will get, as calculated above, after a number of generations a large number of species, and but comparatively few heterozygotes.

The effect of selection can therefore be nothing else than a reduction of the number of species arisen, by elimination of the ones less well fitted to survive, than those which are not eliminated.

Now this is exactly what happens.

About the effect of cold as a selector Nilsson-Ehle says in the article quoted above:

Besonders schnell degeneriren im Klima Schwedens wenig winterfeste Sorten (Linneons according to our nomenclature); die abweichenden Typen nehmen, wenn sie winterfester als die echten sind, nach strengen Wintern sehr rasch an Zahl zu und können bald vorherrschend werden. Die Sorte verliert dann zulezt gänzlich ihren ursprünglichen Charakter."

Now what can be expected to happen in a freely intercrossing community, say if the dominant type is for some reason or other favored through extermination of the recessives in proportionally greater numbers than the dominants?

Such a selection must necessarily also favor the hybrids with dominant appearance because, if f. i. a grey rabbit is favored, on account of its protective color, such a protection is not the outcome of its genotypical constitution but of its phenotypical aspect, and as this does not differ from that of the pure dominants, pure dominants and dominant-hybrids will profit equally by their protective coloring.

Suppose we have an F_2 generation consisting of 1 AA plant 2 Aa plants and 1 aa plant and of these aa Plants 50 % perish on account of their conspicuousness, then we will have 1 AA plant 2 Aa plants and ½ aa plant left. If we assume now, as in the example on p 76 that each plant forms 4 gametes, we get 4 A gametes from the AA plant, 8 gametes from the two Aa plants of which 4 are A and 4 a, and 2 a gametes from the ½ aa plant, which will combine:

$$
\begin{array}{llll}
8\ \text{A} & 6\ \text{a} \\
8\ \text{A} & 6\ \text{a} \\
\hline
& & \times \\
64\ \text{AA} & 48\ \text{Aa} \\
& 48\ \text{Aa} & 36\ \text{aa} \\
\hline
64\ \text{AA} & 96\ \text{Aa} & 36\ \text{aa}
\end{array}
$$

consequently in the proportion 16 AA : 24 Aa : 9 aa against the proportion 16 AA : 32 Aa : 16 aa in the preceding generation.

Or, if we count AA and Aa together as phenotypically indistinctible, we get the proportion 40 : 9 against 48 : 16 or 4⁴⁄₉ : 1 against 3 : 1. But not only has the number of phenotypically dominant forms increased considerably, as compared with the recessives, but also the number of pure dominants, as compared with the number of dominant-hybrids, this being in the preceding generation 16 : 32 or 1 : 2 and in this generation 16 : 24 or 2 : 3 in other words an increase from ½ to ⅔.

6

Consequently: f a v o r i n g   o f   t h e   d o m i -
n a n t s *in a freely intercrossing community tends to
cause an overwhelming majority of the dominant form,
simulating uniformity pretty soon, because the dominant
hybrids are indistinctible from the pure dominants.*

And this is exactly what we see in Linneons like
wild rabbits, where free intercrossing reigns supreme,
thus giving great support to Darwin's idea of the great
influence of selection in nature. Furthermore, if it is
true, that there are no strict selffertilizers in nature, we
must find it easier to obtain, even in so called selffertilizers, the recessives pure than the dominants, because
then, there must be among the phenotypically dominant forms some dominant-hybrids.

Now this is exactly what does occur. About it Nilsson Ehle says in his article, quoted already:

„Bei der praktischen Reinhaltung van Weizensor-
„ten mittels Entfernung abweichender Aehrentypen
„aus den Vermehrungen spielen ferner die Merkma-
„le der Weizensorte ob dominant oder recessif eine
„sehr wichtige Rolle. Eine Sorte mit kahlen oder
„weissen Aehren (recessive Merkmale) ist leichter von
„behaarten bezw. braunen Abweichungen rein zu
„halten als eine behaarte oder braune Sorte (domini-
„rende Merkmale) von Kahlen bezw. Weissen. Bei der
„Entfernung der Abweichungen werden nämlich im
„vorigen Falle auch eventuelle Heterozygoten mit ausge-
„zogen. Im letzten Falle bleiben aber die Heterozygo-
„ten zurück, wenigstens zum Teil, weil sie, ebenso wie
„die typischen Individuen, braun bezw. behaart sind
„und spalten deshalb im nächsten Jahre wieder weisse

„bezw. kahle Individuen ab; auch nach sorgfältig-
„ster Reinpflückung erhält man somit das nächste Jahr
„wieder dieselben Abweichungen, obwohl eventuell in
„geringerer Zahl."

In cases where crossing is entirely excluded, the
only selection possible is through extermination of certain species within the Linneon. That, although strict
fertilizers may not exist, at least some forms cross
much less frequently than others, results from Nilsson-
Ehle's experience in a mixed planting of Pudelweizen
and a wheat designated as n°. 0728, which showed
that the first kind was spontaneously crossed by the
latter, but the latter not by the first.

Yet, so great a selfprotection against crossing is rare
in nature.

So that, if our contention is correct, that a Linneon is
a group of individuals of different constitutions, which is
but a rest of many more different types which sprang
from a cross, and if selection causes a freely intercrossing
community in nature to become apparently uniform
in aspect, we may expect to find that, especially
among animals, the Linneons will, by their uniform
aspect, tend to simulate species.

This question is of such considerable importance,
and touches so closely the question how one came to
consider the appearance of individuals, deviating from
the form, considered to be typical for a particular
Linneon, as evidence of variability, that I may be
permitted to enlarge a little upon it.

Let us take as an example the wild rabbits, referred to the Linneon: Lepus cuniculus.

Every child will tell you: wild rabbits are grey, but people better acquainted with rabbits, gamekeepers f. i., know that occasionally black and orange-yellow rabbits occur in nature. As these are evidently much in the minority compared with the grey ones, they jump to the conclusion that the grey ones are the original form, while the black and the yellow ones are the *varieties* derived from this original form.

Systematists use to express this by saying that the most common form is the „species", while the rare ones are „varieties".

Now the cause of this conception really is nothing but the overwhelming majority of the greys.

If $^1/_3$ of all wild rabbits were grey, $^1/_3$ yellow and $^1/_3$ black, one would probably have made 3 species of them, and never have entertained the idea of variability or, if yet some systematist had united them to one species, one would anyhow, if the idea of a common descent had cropped up, been at a perfect loss to decide which form was the original one: the grey, the black or the yellow one, which had to be considered to be derived.

So that, if it was not exclusively the fact that within most Linnean species, especially within animal ones, one form is in the overwhelming majority which caused the conception of variability, it certainly was this fact which led to the conception of a *single* form being the ancestor of all others.

Now as we calculated already, there is no reason whatever for such a contention.

By crossing the type in the majority with those in the minority, we find that the type in the majority is the dominant one, the others prove to be recessives in regard to this type.

And as we find always that the type in the majority in nature, *is* the dominant one, this dominant type has proved — on our contention that its frequency is due to selection of the most resistant type — to be stronger than the recessives.

Now although „most resistant" and „strongest" are rather vague expressions, we all know that when we are dealing with organisms from the same region in which our experiment gardens are situated, — which is essential for comparison, because what is most resistant in our climate may prove to be least resistant say in the tropics — the recessives usually are by far the weaker forms.

And this also supports Darwin's contention of considerable selection taking place in nature.

But.... the effect of changed conditions may also be the survival of recessives.

We saw that free intercrossing within the Linneon in nature, must finally cause an impression of uniformity, because the hybrids are indistinctible from the favored dominant form and thus are mistaken for the latter. The hybrids among the apparently pure dominants throw recessives only, if they pair with another hybrid, while if they pair with a pure dominant, all children have again the aspect of the dominants. Even without selection this causes, as we all know, already a considerable majority of the type exhibiting the dominant cahracters, compared to the frequency in which the recessive type occurs; this majority being in F_2 of a cross of two forms differing in one respect only, 3 : 1. If, as in rabbits, the dominant colour f. i. is selected for the protective effect it confers on the animals possessing it, this majority of course increases rapidly.

We can imagine however easily, that a change of conditions f. i. a migration of the rabbits from the dunes to black earth, would divest the dominant color, grey, from its protective power, and thus cause an increase of the, now protectively colored, black recescives.

This might even go so far that only one type, recessive in all characters, survives in which case the Linneon has been reduced to a pure species and consequently stops to „vary."

As „variation" by crossing is — by the plasticity it confers on the Linneon — a means and probably the chief one, to adapt such a Linneon to changing conditions the dying out of certain Linneons may have been caused by such a survival of pure recessives only, putting a final stop to the possibility of such a species to adapt itself to changed conditions, unless a cross with an individual belonging to another Linneon, creates fresh opportunities to „vary."

If no such cross occurs, if such a Linneon reduced to a pure species is doomed to remain single, it is — if conditions change — unable to form a progeny able to adapt itself to these changed conditions, and consequently must die out, exactly as human families whose members remain single; though it may remain in existence for a considerable time, if no untoward

change in the conditions to which it is itself adapted
or, if one prefers, resistant, takes place. While thus
extermination of dominants may lead, by the narrowing
down of the Linneon to a single recessive species, to
its extinction, elimination of the recessives never leads
to their total extinction, because these recesives
survive cryptomerously in the gametes of those hybrids which are indistinctible from the dominants,
and therefore share all protection which the latter
may enjoy.

The idea that the aberrant types — the recessives —
are varieties of the most common type — the dominant one — is consequently a mistake; the aberrant
forms never arose from the dominant one, but are segregates from the hybrids indistinctible from the dominant type. Hereditable variability
spells segregation.

The Linneon: Lepus cuniculus, consequently is not
a unit but a group of different types and their hybrids.

That this is really the case would be difficult to prove in the case of the wild rabbit, because we can get
this proof only by mating two hybrid grey forms and, as
these are indistinctible from the pure dominants, a
very large number of matings would be required to
find a pair of such hybrids.

If we disposed of a sufficiently large number of
pure black wild rabbits, it would however be comparatively easy to get the impure grey males because a grey
male mated with such a black, giving grey children
only, would be pure; a grey male mated with such a
black (preferably with the same), giving a mixture of

black and grey children, on the other hand, would betray its hybrid nature.

In the same manner mating grey females with a black male would reveal to us the purity or impurity of the grey female which, if pure, would get grey children only, if impure, grey *and* black ones. An impure grey male and an impure grey female, so found out, paired together, would give a certain percentage of black children and thus prove our contention that there must be hybrids among the apparently pure wild grey rabbits.

The quickest way to obtain such impure greys would probably be to catch a pregnant black female, as such a one is very likely to owe its pregnancy to a grey male, pure or impure, and consequently will either throw a litter of exclusively impure greys or consisting of a mixture of impure greys and pure blacks.

The fact that in Java frequently black and spotted panthers are found in the same litter, is conclusive proof that among either the black or the spotted panther in Java, there are hybrid individuals.

All such kinds of impurity are much easier demonstrated in the case of plants which can be selffertilized, because here isolation of a sufficiently large number of individuals of the phenotypically dominant type reveals to us, by their several progenies, at once the presence of hybrids.

In this way, I was able to show that among our ordinary brown wallflowers (Cheiranthus Cheiri) some are pure dominants, while others are hybrids, segregating violets, golden-yellows and whites, thus showing that the wallflower is not a species but a Linneon, con-

sisting of four different species: brown—violet—gol-
denyellow—and white-flowering ones and their hybrids.

With this demonstration of course the real origin of
these species is not explained.

It probably lies, as experiments with Antirrhinums
show, in a previous cross between two individuals
belonging to different Linneons.

Such a cross gives us — if the forms originally crossed were pure — a uniform F_1, each individual of
which, after selfing, gives rise to a large number of differently constituted individuals, some of which are heterozygotes while others are homozygotes or pure species.

Every cross between two individuals, differing in
many respects, consequently gives rise to the production of a mass of differently constituted types, far greater than the number of separate types usually found
within a wild Linneon.

This is caused by the fact that of the number of
different forms which were born from a cross, only a
small number survives.

The Linneon is consequently the
vestige, *consisting of a comparatively small number of differently constituted types,* of that very
large number of types born as the
result of a cross.

The number of types born from crossing in nature,
is usually even much larger than that resulting from
a cross we effect purposely in our experimentgarden,
because in nature crossing is not limited to a single initial cross, but usually a large number of crosses occurs
at the same time. This can be esasily demonstrated by

planting a single individual of Antirrhinum glutino-sum in an experimentgarden, in which a large number of different types of Antirrhinum are cultivated.

As Antirrhinum glutinosum is selfsterile, we can observe the result at once in F_1.

Now the result, obtained in such a case in Bennebroek, was an enormunsly varied F_1 which could be explained, either by the fact, that the individual of A. glutino-sum had been crossfertilized by a number of different Antirrhinum-types, the pollen of which had been carried by humblebees to the different flowers of the glutinosum-individual, or by a single, but highly heterozygous, type.

A priori, the former was the most probable but this single experiment was insufficient to settle what had really happened.

To settle this point, several individuals of A-glutino-sum were planted out and seeds of one of them collected.

The result was the same: a multiform F_1, but this time there were pure glutinosum-forms among this F_1, which failed to appear in the first one.

It was thus proved that A. glutinosum was fertilised at least by two of the different types present in the garden viz by A. glutinosum *and* by at least one heterozygous type of A. majus, much more probably by several types, the pollen of which had been brought to the glutinosum plant by insects.

This experiment is yet in progress with the overwintered F_1 plants which of course will continue to cross with one another and with other types present in the garden, thus giving continuously rise to new types,

just as will happen in nature, and so continue to cre-
ate yearly a number of new types which are able to
fill a good many opportunities for existence, present in
their neighbourhood or within a distance to which
their seeds may be carried, thus offering in other
words, splendid material for adaptation.

What is born, consequently depends on what is crossed,
what survives, on the local circumstances of the birth-place
of the new forms and on those of its neighbourhood.

If the same kind of cross takes place in Japan and in
Holland, the same forms will be born at those two
distant places. So I found f. i. a white Mus rattus
with a brown head and brown shoulders caught in
Leiden, hardly distinctible from a similar form from
Japan in the Leiden Museum .

As the Linneon Mus rattus ranges from Holland to
Japan, it evidently gives rise to the same recessive
forms, when two hybrids indistinctible from the black
dominants pair, irrespective of the spot where this
pairing takes place.

Now it might very well happen, that the black form
disappeared from Japan and survived in Holland,
through local conditions favoring the survival of the
recessive form in Japan, in which case these two forms
would be considered as different „species", in support
of which distinction, one would not fail to lay stress upon
their occurence in very distant, not overlapping areas.

Such a thing will happen especially frequently with
Linneons of plants, containing forms of a different degree of resistancy against frost which leads to a difference in the composition of such Linneons in different

countries, causing them to be considered „specifically" different. That this *is* so, is shown by Nilsson-
Ehle's experience that certain „species" of wheat, introduced into Sweden, and cultivated there for some
time, become of quite another type by the weeding
out of the but little frostresistant strains by the
Swedish climate.

In many cases, as we saw, freely intercrossing Jordanons cause the Linneons to get, in nature, a uniform aspect.

Yet there are cases in which such Linneons show a
bewildering multiformity either in several or in one
respect, even at first sight.

The most beautiful exemple of this, I ever saw, is furnished by a series of about 200 specimens of Buteo
Buteo in the Leiden Museum, hardly two of which are
alike, and which resemble very much, what one would
expect to obtain, after segregation in F_2 [1]).

The reason, that this lot resembles an F_2 generation
probably is that here *no selection* has been at work,
because this bird of prey is so strong that it has
practically no ennemies in the regions in which it occurs.

In cases, where the Linneon is more uniform in aspect but yet shows great diversity in one respect, the
explanation probably is, that as far as the „variable"
character is concerned, selection did not take place.

So f. i. Symphytum officinale which, at least in the
neighbour hood of Bennebroek is self-sterile and consequently forms an intercrossing community, like that

---

[1]) I regret it greatly that it is not possible to publish a colored plate
of this most important series.

of Lepus or Buteo, shows all kinds of colors in its flowers, f. i. dark-violet, light violet, different shades of red, whites etc.

The same is the case with the blue-eyed, grey-eyed, brown eyed and black eyed human individuals.

The explanation is probably that none of these colors gives a special advantage in our region and consequently there is no elimination.

That, notwithstanding these differences, we continue to consider such heterogenous groups as Linneons, is because we abstract from differences, considered to be unessential, such as color of flowers etc. and take so called essential characters as criteria as explained on p. 50.

That other self-sterile forms behave just the same among plants as among animals, e. g. tend to become uniform by selection of certain characteristics, is proved by the behaviour of the strictly-self sterile Cardamine pratensis inside of which Linneon no such striking differences in the colors of the flowers occur.

In other cases, in which we find great diversity of form, as f. i. within the Linneon: Taraxacum vulgare, the cause is very different from that which causes diversity in the case of Symphytum.

Taraxacum vulgare is — with the exception of one form in Japan —, notwithstanding the fact that its flowers are perfect and produce quantities of pollen, never fertilized, but reproduces itself apogamously; consequently selection can have no other effect than to decrease the number of different forms, just as in the case of strict self-fertilizers.

Other polymorphous Linneons, like some in the genera Hieracium and Antennaria, though occasionally able to reproduce themselves in the normal sexual way, and consequently open to an occasional cross, yet usually, behave like Taraxum which explains their high degree of polymorphism.

Of course, if apogamy follows directly upon a cross, it perpetuates not only the homozygous forms but the heterozygous ones as well — just as cuttings do — so that the conclusion, drawn by EAST AND HAYES from the fact (if it be a fact) that such Hieracia occasionally „vary" and consequently crossing can not be the only cause of „variability", is unwarranted because of course heterozygous apogamous forms can give rise to a number of different forms, can „vary" as a result of vegetative segregation, shown by East himself, to exist f. i. among potatoes.

Summa Summarum, I think we may say *that a Linneon is a vestigial group of a once much larger group of differently constituted types, born from a cross, which is apt to simulate a species by the overwhelming majority of the dominant types it contains, as a result of free-intercrossing, combined with a favoring of the dominants by a process of selection, weeding out the weaker or more conspicuous recessives; this uniformity being more apparent than real, because pure dominants are indistinctible, in most cases, from dominant-hybrids.*

A Linneon consequently is nothing but a group of morphologically *similar* individuals. It may consist of almost nothing but pure species and a few hybrids, as it does in habitual self-fertilizers; it may consist of a mixture of homozygotes and heterozygotes, reprodu-

cing themselves apagomously as in the case of Taraxum officinale, it may consist chiefly of pure dominants and hybrids, resembling these, with a slight ad-
mixture of pure species (the recessives) as in the case of
the wild rabbits or of the brown-flowered Cheiranthus
Cheiri, or it may consist of almost nothing but hybrids of different constitutions, as it does f. i. in the
case of Symphytum officinale and of human beings.

That, even in such cases as these last ones, we keep up
the conception of a Linneon, is caused by the fact that
we are impregnated, from infancy almost, with the conviction that some characters are essential for a Linneon
while others are not, so that we refuse to cut up a Linneon into smaller groups, as long as these „essential"
characters e. g. those common to all the different types
within the Linneon, are not brought into play, which
of course they never can, as we ourselves did limit the
Linneon by the criterium of the very presence of these
characters in all its individuals.

CHAPTER VIII.

THE LIMITS OF LINNEONS.

That Linneons are more or less distinctly limited, is apparent from the fact that in subdivising the living Kingdom, one made unconsciously halt at the border of each Linneon, and so was led to consider them as real species, such as Linnaeus believed them to be.

In most cases the limits between the smaller units within the Linneon, are much less conspicuous, so that it lasted until the middle of the 19th century before Jordan discovered even the mere existence of such smaller units within the Linneon.

This difference in the distinctness of the limits between the Jordanons and the Linneons needs explaining.

We must enquire into this question by asking: are there cases, in which the Jordanons within a Linneon are as distinct as the Linneons themselves and are there cases in which this is not the case?

The answer is that there are, so that the question remains, what causes this difference?

Investigation shows that poor distinctness of the Jordanons within the Linneon, occurs in all cases where these Jordanons intercross freely and selection favors a certain type, and that this distinctness increases in the same ratio as intercrossing decreases, so that it reaches its maximum when crossing does not occur at all, as in those cases

in which the different types are perpetuated apogamously.

*This gives us the clue as to the difference in distinctness
of the units within the Linneon and between the different
Linneons themselves; the units within each Linneon form
an intercrossing community, while the Linneons themselves usually do not intercross.*

Why not?

Because mostly, individuals belonging to different
Linneons, have either an aversion to mating or are prevented from mating, by isolation either in space (occurrence in different regions) or in time (different time of
flowering or different time of rutting) or by mechanic
obstacles (differences in size of the male of the one and
the female of the other Linneon, non-fitting copula-
ting organs etc.), all this accentuated by the fact that in
many cases, even if mating occurs, no progeny or a
sterile progeny only, results.

*The distinctness of the Linneons is consequently caused
by the obstacles against mating of the individuals belonging to different Linneons, which obstacles may be relative obstacles which can be overcome, such as aversion f. i.
or final obstacles which cannot be overcome: innate sterility as f. i. between many Linneons of the genus Verbas-
cum .*

*Consequently it is nature itself which groups the indi-
viduals to Linneons and Linneons are thus something
more than mere conceptions of the human mind; it are
natural intercrossing communities of differently constitu-*

7

ted types. As we have seen that constant intercrossing, such as takes place inside of many Linneons between the different types, by the selection of a certain type, which is the rule in nature, has finally a swamping effect, leading to an overwhelming majority of the dominant type — including both pure dominants and hybrids indistinctible from these at sight — there is no reason to suppose that what happens within the Linneon would not happen between the several Linneons themselves, if they also intercrossed freely, so that it is reasonable to suppose that if there were no obstacles of any kind to a free intercrossing in nature between all the differently constituted types, which people the earth, this latter would be chiefly peopled by one type only.

The cause of the possibility of a great diversity of types, living side by side in the same regions, is the existence of obstacles to free intercrossing, isolating these apparently non-isolated types as effectively as if the different types were put into separate cages.

Within each cage intercrossing freely occurs and leads, by the aid of selection, to a high degree of phenotypical uniformity, while the numerous phenotypically different types so obtained — the Linneons — remain distinct because the walls of the cages — in nature the obstacles to crossing — keep them separate.

Linneons consequently, though being themselves the vestiges of the result of a cross, are kept distinct in nature by obstacles against their freely crossing with other Linneons.

If there existed no obstacles to unlimited intercrossing in nature, we would be unable to distinguish Linneons; if no crossing took place at all, Linneons—being themselves the result of a cross — would never have originated, so that *Linneons owe their origin to the occasional possibility of a cross and their persistence to the bars which nature, as a rule, keeps closed to prevent intercrossing, and but occasionally opens.*

The view, that distinctness of the Linneons is caused by bars to intercrossing with other Linneons, is greatly supported by the fact, that in cases as those of the Willows where the Linneons are, as we all know, very badly limited, this bar against intercrosssing with other Linneons does not only not exist, but such intercrossing is, on the contrary, favored (see Chapter XI).

Whenever nature allows crossing, evolution sets in; whenever crossing is made impossible it stops as soon as the segregation initiated by the cross, is at an end.

As the bars, separating the Jordanons within the Linneons, are usually open, evolution within the Linneon, the formation of so called new „varieties" is a phenomenon occuring daily, and it is stopped permanently only by the cropping up of apogamy.

As on the contrary, the bars between the different Linneons are usually closed, the origin of a new Linnean species is a phenomenon of much rarer occurrence and very frequently stopped definitely by the innate sterility of two Linneons.

The production of new Linneons, usually called „species", is generally designated as progressive evolution, the production of new Jordanons, usually called „varieties" is frequently called degressive evolution.

———

CHAPTER IX.

THE CAUSE OF THE INCREASE OF „VARIABILITY" UNDER DOMESTICATION.

It is almost generally believed that domestication causes variation by the influence of better food or of unusual food etc., giving rise to certain transmittable tendencies in the domesticated animals or plants.

So Darwin says:

„When we see an animal highly kept, producing off-
„spring with an hereditary tendency to early maturity
„and fatness, when we see the wild duck, and the austra-
„lian dog, always becoming, when bred for one or a
„few generations in confinement, mottled in their co-
„lours .... we naturally attribute such changes to the
„direct effect of known or unknown agencies acting for
„one or more generations on the parents. It is possible
„that a multitude of peculiarities may thus be caused
„by unknown external agencies".

Now we have seen that there is no proof for the existence of a transmittable influence of external agencies, so that we must look for another cause for the changes we find that frequently follow upon confinement of a wild animal. Darwin himself clearly recognized that there must be other causes for the greater „variability" existing among domesticated animals and plants, than this direct effect of external agencies, so that he ascribes a much greater effect to what he calls *indirect agencies*:

„I may add, judging from the vast number of new varieties of plants which have been produced in the „same districts and under nearly the same routine of „culture, that probably the indirect effects of domes-„tication in making the organisation plastic is a much „more efficient source of variation than any direct „effect which external causes may have on the colour, „texture or form of each part".

Now what may this *indirect* effect of domestication be?

To answer this we must first ask: *is the fact that we* s e e *a larger number of different forms belonging to the same Linneon under domestication than in nature, proof that there* e x i s t *more such forms under domestication than in nature ?*

This of course n e e d not be the case, it is also possible that forms are visible under domestication which, although existing in nature, remain there hidden to us.

Suppose this were the case, what then would cause their appearance under domestication?

The answer is: isolation.

We have seen that every heterozygote isolated, undergoes segregation by which the recessives it contains in a cryptomerous way, become visible.

Now this is exactly what domestication does: it isolates individuals, and this simple fact explains how savages, of whom no great „breeding" qualities can be expected, yet succeed in raising different races of domesticated animals and plants.

In plants this succeeds, as we saw from the example of the wallflowers, easily.

Suppose a savage takes home for adornment of his garden, a wild brown wallflower. Planted, this individual, if heterozygous, will, from the seeds it scathers around it, raise an offspring among which there will be some with white flowers.

Now suppose the savage prefers this white-flowered race above the brown-flowered one, and consequently pulls out all brown-flowered plants which sprang up in his garden, he succeeds at once in obtaining this white flowered form — which happens to be the recessive — pure, and has thus, with very little trouble or insight, obtained a new constant race.

In the same way, it is easy to obtain new races of rabbits. Suppose a heterozygous pregnant wild rabbit has been caught by a savage and put into a cage, and let us suppose further — taking the most unfavorable example — that this heterozygote was pregnant from a pure grey male.

The litter thrown, will then consist of pure greys only, but as some of the males in this litter will be heterozygotes, such a heterozygous male will, paired with its mother or with a heterozygous sister, give some aberrant offspring by segregation, which aberrant forms, if bred together, will give easily rise to new races.

Isolation and subsequent selection of the aberrant types consequently suffises to obtain the aberrant forms cryptomerously hidden in wild animals and plants, and to breed these true to type. This, in all probability, has

*almost unconsciously been done in all efforts to tame
wild animals or to cultivate wild plants.*

Domestication, even without introducing a new source for the production of new forms, consequently
allows us to gain forms, not or but
very rarely met with in nature, by
the mere isolation of heterozygotes.

But domestication has not acted in this way only; it
certainly has introduced new sources for the production of new forms by crossing.

It is a very curious fact, that this evident source of
„variability" under domestication has always been
explained away.

Almost all writers on domestication of animals or
plants had to acknowledge that much pointed towards
a multiple origin of our domesticated races by crossing,
and yet almost all have, notwithstanding this evidence, pleaded for a single origin by variation from one
ancestral form. |

We all know, that Darwin ascribed the origin of the
domesticated as well as of the wild new forms to some
sort of heriditary variation. For this conception it is
evidently necessary, to show or at least to make plau-
sible, that the domestic races, as well as the wild „varieties", can be considered to belong to one species.

But if we examine the evidence for this contention,
we at once perceive how meagre it is.

In support of this, I will quote from Darwins „Variation of animals and plants under domestication second
edition revised. London John Murray 1893.

Domestic dogs.

Vol. I, p. 26: „it is highly probable that the domestic dogs of the world are descended from two well-
„defined species of wolf (viz C. lupus and C. latrans)
„and from two or three doubtfull species (namely the
„European, Indian and North African Wolves);
„from at least one or two South American canine spe-
„cies; from several races or species of jackal, and per
„haps from one or more extinct species".

Domestic Cats.

Vol. I, p. 49. „we have seen that distant countries
„possess distinct domestic races of the cat. The diffe-
„rences may in part be due to descent from several
„original species, or at least from crosses with them".

Domestic Horses.

Vol. I, p. 53. „Whether the whole amount of diffe-
„rence between the various breeds has arisen under
„domestication is doubtfull. From the fertility of the
„most distinct breeds, when crossed, naturalists
„have generally looked at all the breeds as having
„descended from a single species. Few will agree with
„Colonel H. Smith, who believes that they have des-
„cended from no less than five primitive and diffe-
„rently coloured stocks. But as several species and
„varieties of the horse existed during the later ter-
„tiary periods and as Rütimeyer found differences in
„the size and form of the skull in the earliest known
„domesticated horses, we ought not to feel sure
that „all our horses are descended from a single
species".

The ass.

Vol. I, p. 65. „There is now little doubt that our „domesticated animal is descended from the Equus „taeniopus of Abyssinia.

Pigs.

Darwin supposes our Pigs to have arisen from crossing Sus indicus and Sus scrofa and says:

Vol. I, p. 74. „Seeing how different the chinese pigs, „belonging to the Sus indicus type, are in their os „teological characters and in external appearance „from the pigs of the S. scrofa type, so that they „must be considered specifically distinct,, it is a fact, „well deserving attention, that Chinese and common „pigs have repeatedly been crossed in various man „ners with unimpaired fertility".

Cattle.

Vol. I, p. 82. „Domestic cattle are certainly descen „dants of more than one wild form, in the same man „ner as has been shown to be the case with our dogs „and pigs."

Sheep.

Vol. I, p. 97. „Most authors look at our domestic „sheep as descended from several distinct species." p. 98. „Another ingenious observer though not a „naturalist, with a bold defiance of everything known „on geographical distribution, infers that the sheep „of Great Brittain alone are the descendants of eleven „endemic British forms! Under such a hopeless state „of doubt it would be useless for my purpose to give „a detailed account of the several breeds."

Goats.

Vol. I, p. 105. From the recent researches of M.

„Brandt most naturalists now believe that all our
„goats are descended from the Capra aegagrus of
„the mountains of Asia, possibly mingled with the
„allied species C. falconeri of India"

*Rabbit*s.

Vol I p. 107 „All naturalists, with as far as I know
„a single exception, believe that the several domestic
„breeds of the rabbit are descended from the com-
„mon wild species; I shall therefore describe them
„more carefully than the previous cases."

Pigeons.

Vol. I, p. 137.   „I have been lead to study domestic
„pigeons with particular care ,because the evidence
„that all the domestic races are descended from one
„known source is far clearer than with any other
„anciently domesticated animal".

Fowls.

Vol. I, p. 251. „Finally we have not such good evi-
„dence with fowls as with pigeons, of all the breeds
„having descended from a single primitive stock."

Ducks.

Vol. I, p. 295. „From these several facts, more espe-
„cially from the drakes of all the breeds having cur-
„led tail-feathers and from certain sub-varieties in
„each breed occasionnally resembling in general
„plumage the wild duck, we may conclude with
„confidence that all the breeds are descended from
„Anas boschas."

The Goose.

Vol. I, p. 302. „A large majority of capable judges
„are convinced that our geese are descended from

„the wild Grey-leg goose (A. ferus); the young of
„which can easily be tamed".

The Peacock.

Vol. I, p. 305. „This is another bird, which has hard-
„ly varied under domestication, except in someti-
„mes being white or piebald. Mr. Waterhouse care-
„fully compared, as he informs me, skins of the wild
„Indian and domestic bird and they were identical
„in every respect, except that the plumage of the
„latter was perhaps rather thicker."

The Turkey.

Vol. I, p. 308. „It seems fairly well established by
„Mr. Gould, that the turkey, in accordance with the
„history of its first introduction is descended from
„a wild Mexican form which had been domesticated
„by the natives before the discovery of America, and
„which is now generally ranked as a local race and
„not as a distinct species."

The Guinea-fowl.

Vol. I, p. 310. . . . . „is now believed by some natura-
„lists to be descended from the Numida ptilorhynca,
„which inhabits very hot, and, in parts, extremely
„arid districts in Eastern Africa. *Consequently it has*
„*been exposed in this country to extremely different*
„*conditions in life. Nevertheless it has hardly varied*
„*at all except in the plumage being either paler or*
„*darker colored"* [1]).

The Canary-bird.

Vol. I, p. 311. „It has been crossed with nine or ten

---

[1]) Italics are mine.

„other species of Fringillidae, and some of the hy-
„brids are almost completely fertile; but we have no
„evidence that any distinct breed has originated
„from such crosses.

Gold-Fish, Hive-bees and Silk-moths. No definite state-
ments.

Plants.

Cereal plants.

Vol. I, p. 338. „Finally, every one must judge for
„himself whether it is more probable that theseveral
„forms of wheat, barley, rye and oats are descended
„from between ten and fifteen species, most of which
„are now either unknown or extinct, or whether they
„are descended from between four and eight species
„which may have either closely resembled our pre-
„sent cultivated forms, or have been so widely diffe-
„rent as to escape identification."

Zea Mays.

Vol. I, p. 338. „Botanists are nearly unanimous that
all the cultivated kinds belong to the same species".

Cabbage.

Vol. I, p. 343. „Most authors believe that all the ra-
„ces are descended from the wild cabbage found on
„the Western shores of Europe, but Alph. de Can-
„dolle forcibly argues, on historical and other grounds,
„that it is more probable that two or three closely
„allied forms, generally ranked as distinct species, still
„living in the mediterranean region, are the parents,
„now all mingled together, of the various cultivated
„forms".

p. 344. „The other cultivated forms of the genus

„Brassica are descended, according to the view a-
„dopted by Godron and Mezger, from two species, B.
„napus and B. rapa; but according to other bota-
„nists from three species, whilst others again strong-
„ly suspect that all these forms, both wild and cul-
„tivated ought to be ranked as a single species.
„Brassica napus has given rise to two large groups,
„namely swedish turnips (believed to be of hybrid
„origin); and Colzas the seeds of which yield oil.
„Brassica rapa (of Koch) has also given rise to two
„races, namely common turnips and the oil-giving
„rape."

Peas.

Vol. I, p. 349. „Whether many of the new varieties
„which incessantly appear are due to such occasio-
„nal and accidental crosses I do not know."

Beans.

Vol. I, p. 349. „With respect to beans (Faba vulgu-
„ris), I will say but little.... As in the case of the
„pea, our existing varieties were preceded during
„the Bronze age in Switzerland by a peculiar and
„now extinct variety producing very small beans."

Potato (Solanum tuberosum).

Vol. I, p. 350. „There is little doubt about the paren-
„tage of this plant, for the cultivated varieties differ
„extremely little in general appearance from the wild
„species which can be recognized in its native land
„at the first glance.

The Vine (Vitis vinifera).

Vol. I, p. 352. „The best authorities consider all our
„grapes as the descendants of one species which

„now grows wild in Western Asia.... some authors
„however, entertain much doubt about the single
„parentage of our cultivated varieties".

White Mulberry (Morus albus).

Vol. I, p. 354. „In India the mulberry has also
„given rise to many varieties. The Indian form is
„thought bij many botanists to be a distinct spe-
„cies, but as Royle remarks, so many varieties
„have been produced by cultivation that it is dif-
„ficult to ascertain whether they all belong to one
„species.

The orange group.

Vol. I, p. 355. „We here meet with great confusion
„in the specific distinction and parentage of the
„several kinds."

Peach and Nectarine (Amygdalus persica).

Vol. I, p. 360. „Whether or not the peach has procee-
„ded from the almond, it has certainly given rise to
„nectarines."

Apricot (Prunus armeniaca).

Vol. I, p. 365. „It is commonly admitted that the
„tree is descended from a single species, now found
„wild in the Caucasion region."

Plums (Prunus insititia).

Vol. I, 366. „Formerly the sloe, P. spinosa, was
„thought to be the parent of all our plums but now
„this honour is very commonly accorded to P. in-
„sititia or the bullace, which is found wild in the
„Caucasus and N. Western India.... another sup-
„posed parent-form, the P. domestica is said to be
„found wild in the region of the Caucasus."

Cherries (Prunus cerasus, avium etc.).

Vol. I, p. 368. „Botanists believe that our cultivated „cherries are descended from one, two, four or even „more wild stocks".

Apple (Pyrus malus).

Vol. I, p. 369. „The one source of doubt felt by bo-„tanists with respect to the parentage of the apple, „is whether, besides P. malus, two or three other „closely allied wild forms, namely P. acerba and „praecox or paradisiaca, do not deserve to be ranked „as distinct species".

Pears (Pyrus communis).

Vol. I, p. 372. „I need say little on this fruit, which „varies much in the wild state, and to an extraor-„dinary degree, when cultivated, in its fruit, flowers „and foliage. One of the most celebrated botanists „in Europe, M. Decaisne has carefully studied the „many varieties; although he formerly believed that „they were derived from more than one species, he „now thinks that they all belong to one."

Strawberries (Fragaria).

Vol I, p. 374. „The blending together of two or more „aboriginal forms, which there is every reason to be-„lieve has occurred with some of our anciently cultiva-„ted productions, we see now actually occurring with „our strawberries," [1])

Gooseberry (Ribes grossularia).

Vol. I, p. 376. „No one, I believe has hitherto doub-„ted that all the cultivated kinds are sprung from „the wild plant bearing the name, which is common

---

[1]) Italics are mine.

„in central and Northern Europe."
Walnut (Juglans regia).

Vol. I, p. 379. Description of different „varieties" only.
Nuts (Corylus avellana).

Vol. I, p. 379. Most botaniss rank all the varieties under the same species, the common wild nut.
Cucurbitaceous plants.

Vol. I, p. 384. „Finally M. Naudin remarks that the „extraordinary production of races and varieties by a single species and their permanence, when not interfered with by crossing, are phenomena well calculated to cause reflection.
Trees.

Vol. I, p. 384. „Deserve a passing notice on account „of the numerous varieties which they present."
Flowers.

Vol. I, p. 388. „Many of our favourite kinds in their present state are the descendants of two or more species crossed and commingled together and this circumstance alone would render it difficult to detect the difference due to variation."

As we see, a single origin is *proved* in no case, while an origin from different sources by crossing, is made probable on the other hand in many cases.

Darwin himself resumes the question in the follo- wung sentence on p. 12 of the Origin:

„In the case of most of our anciently domesticated „animals and plants it is not possible to come to any „definite conclusion, whether they are descended from „one or several wild species"

In the face of this fact and of the knowledge, we have obtained, that a transmittable influence of external circumstances has never been proved to exist, as little as any other form of hereditable variability, it is safe to say that:

Our domesticated plants and animals are the results of isolation of heterozygotes, caught in nature, followed by selection and isolation of the recessives, or the results of crossing, followed also by segregation and selection of the desirable segregates.

Now is it possible to get definite proof for this contention?

It seems to me that it is.

If crossing is at the bottom of the origin of our domestic races, such races must have originated as heterozygotes; and as such heterozygotes frequently must have had desirable qualities already, it stands to reason, that in cases, in which such heterozygotes could be multiplied asexually, e. g. by budding, by grafting, by cuttings, by bulbs etc. one will have resorted to one of these means. If therefore, crossing is the final cause of the origin of new domestic races, it is reasonable to expect, that a great majority of the different kinds of trees and flowers, multiplied habitually in an asexual way, as f. i. fruittrees and flower bulbs are, must be heterozygotes.

Now this is doubtless the case.

We know that nearly all kinds of apples, hyacinths, tulips etc. when sown, segregate into a great number of different types — frequently in a most astonishing number of them — thus proving their hybrid origin.

8

We are thus justified to conclude:

Domestication spells segregation, followed by selection and isolation of the desirable segregates.

Crossing is always at the bottom of it, this may have taken place already before the animal or plant was domesticated in nature; in most cases however will have taken place after domesticion.

All breeders of animals and plants know this and continuously obtain novelties by crossing.

The introduction of new forms from distant countries is therefore diligently resorted to, always with the view of crossing them with stock already in hand, in the hope to obtain novelties.

On the other hand, there is not the slightest proof, that the mere change of conditions, following upon domestication, itself causes „variability"; in all known cases such „variability" was the result of a cross.

That it *seems* occasionally, as if spontaneous variability occurs in breeds, is caused by the fact that hardly any breed is homozygotic in all its individuals, so that aberrant types are born, whenever two heterozygotes happen to mate. This f. i. is the reason that from time to time red „variants" occur in the dutch white and black cattle.

CHAPTER X.

PROGRESSION IN EVOLUTION.

We saw in the eighth chapter, that the production of new Linneons, usually called „species", is generally designated as progressive evolution, and that this kind of evolution is considered by many *the* evolution par excellence.

We must therefore devote a chapter to the question of Progression in Evolution.

This question of progression is a vexed one; there is in the human mind a craving towards improvement in everything, which makes mankind believe, but all too readily, that things are better to day than they were yesterday, and wil be better still to morrow.

Perhaps the only good, that will come of the damnable war which is reigning over the greater part of the „civilised" world, will be the recognition that we did not progress as much as we thought we did, that man is little better or rather worse than he was f. i. in roman times.

But man is such a curious animal, that he will forget in a few years how he behaved in these years as a beast, and pride himself again on his „progression"! There is no more conceited being in all nature than man, and this conceit is a snare, in which he is caught every time he looks at the universe from the standpoint, so tickling to his vanity, that he almost unconsciously dwells upon it, that all was made for his pleasure. This anthropocentric standpoint causes him to believe that the world could not possibly have been so good before he was pleased to make his appearance into it, than after this never-sufficiently-to-be-appreciated condescension of his, and so we have had, in all seriousness, a discussion on the supposed inadequacy of the adaptation of the plants of coal measure times.

It is this same anthropocentric pedantry which leads man to consider himself the crown of the universe, and to arrange the different kinds of animals according to their degree of lesser or greater similarity to himself, claiming that those most differing from his Nibbs, are the lower ones, those resembling him most, the higher ones, thus creating the conception of progression.

So a little chit of a monkey becomes „higher" than an elephant, a mouse higher than a Condor, (is not sucking the young, like a human mother, much „higher" than feeding them on carrion) a slow-worm higher than a shark, poor little Amphioxus higher than a giant cuttle-fish, a clam higher than the most beautiful of yelly-fishes and all such nonsense.

Very rightly Victor Franz has maintained in the Biologisches Centralblatt 1911 p. 1, that we possess no reliable measure whatever for the determination of which organisms are higher, which lower on the scale.

The result depends entirely on which characters we choose for comparison. If we choose the brain and the urogenital system as our measure, one can indeed argue that man is the highest animal, but if we choose as criteria the protective measures against wet feet, a cow is an infinitely higher animal than man.

The much met with criterium: higher differentiation indicates a higher, less differentiation a lower degree of development, is no good either; by this measure amphibia, generally considered to be „higher" than fishes, would have to be placed lower than these on the progressive scale, because they lack a good deal of differentation in the skeleton, in the sense-organs of the skin, in the covering of it, in the brains, with the exception of the Pallium, and in the ovarium, which the Teleosts possess.

Comparison of the characters consequently gives no criterium for progression, so that it is very possible that progression has as litte real existence in nature as genera have, but that progression is an abstraction of the human mind just as the genus is.

But.... we are not yet justified to draw this conslusion; there is something, called the geological record, the record — even if it be very incomplete — of the development in time which according to the opinion of many, proves that progression does exist!

It proves nothing of the kind; it proves, at the most, that the succession was: Coelenterata, Molluscs, Fishes, Reptils, Birds, Mammals, to mention only the groups referred to above, but this of course does not prove, that the forms which made their appearance the latest, are the highest ones (did not much „higher" reptils appear much earlier than the present, „lower" ones?). One might, with equal justice, claim mental superiority for the child which is last born!

Consequently the geological record gives no support to progression either, and we are perfectly justified to say that progression is a human conception and that progressive evolution does not exist.

But the geological record does show — even if we take full account of its incompleteness — that man appeared very late on the globe, that fishes and amphibia preceded reptilia and mammalia etc.

This has to be accounted for.

But this is no accounting for p r o g r e s s i v e *evolution, but for* s u c c e s s i v e *evolution, as opposed to simultaneous creation.*

Now this succession shows that the later types have arisen from gametes, produced by the immediately preceding types, and consequently that under the conditions existing and having existed on the earth — in the widest sense — no other way was possible.

But this shows, by no means, that there is something innate to living matter which necessitated this particular sequence; we might very well have had an earlier appearance, say of man, than has taken place, if conditions — in the widest sense — had been different.

This is clear, when we keep in mind that in our experiment gardens we have it in hand, to postpone to any desirable moment, the production of a new species, by not executing the cross, as a result of which, that species would arise, or to hasten its production by executing that particular cross at once.

So it may be, that if the cross from which man has arisen, had been executed by nature earlier than it has been executed, man would have appeared earlier in the history of the earth, than he has. His production may have been, what is usually, called the result of chance e. g. of the meeting and mating of two forms, at that particular moment and not at an earlier one, just as the moment of the birth of children depends, in the last instance, on the chance meeting of the man and woman who, after this meeting, decide to marry.

In how far chance, in this sense, has played a role in successive evolution is an unanswerable question; could man f. i. have originated from gametes, produced by two particular reptils, if these had happened to mate, or is a shortening of the sequence: reptils— „lower" mammals — man, impossible?

We know not; and the so-called Biogenetisches Grundgesetz gives no solution either, although it rather points towards the possibility of shortening.

All this is speculation; we have anyhow not to deal with the question in which other way evolution *might* have taken place, but how it *has* taken place and this we can deduce from the geological record only, which shows clearly that it took place successively.

With such a successive evolution, our theory that new Linneons arise from a cross of preexisting ones, is in full accord, so that the moment has come to consider the evidence for the occurrence of crossing of different Linneons in nature.

CHAPTER XI

THE EVIDENCE FOR THE OCCURENCE OF CROSSES BETWEEN INDIVIDUALS BELONGING TO DIFFERENT LINNEONS IN NATURE.

Crossing of individuals belonging to different Linneons will be brought about by opportunity or by necessity.

Opportunity is offered f. i. whenever pollen gets on a foreign stigma and nothing prevents its subsequent fertilizing action; necessity is born by the sexual desire which no animals and few men can resist in the long run.

So, while most white men in Europe would not think of mating with a negress, white men not infrequently do so in Africa, where no white women live, and while bantamcocks don't pair with pheasants in nature, they do so in captivity, if locked up with pheasant-hens, in the absence of hens of their own kind.

While hare and rabbit don't pair in nature, a male hare doubtless would do so if sufficiently long isolated with female rabbits, in the absence of male rabbits on an island, as ressorts from the experiments of Mr. Houwink, showing that the hare looses its inborn aversion of a tame rabbit, if it is taken soon after birth from its mother, and sucked by a tame rabbit foster-mother.

This secret for obtaining Leporids, Mr. Houwink obtained from a poacher and it worked so splendidly that he is now in the possession of a fine fullgrown F_1 generation of Leporids, a male of which has already paired with one of its sisters. [1])

I would therefore not be a bit surprised if it were subsequently proved that our tame rabbits were evolved from an original cross between the wild rabbit and the hare, which cross Mr. Houwink is trying to obtain in a similar way.

Perhaps this trick of having the individuals belonging to a foreign Linneon nurtured by an individual belonging to the Linneon, with which one desires to cross them, is very widely applicable and may lead to the obtention of many interesting hybrids.

Dr. van Oort tells me that it is well known, that a bitch of our domestic dogs refuses nearly always to mate with a male wolf, but shows no aversion if she herself was raised by a female wolf.

The aversion to mating between individuals belonging to different Linneons can consequently be overcome, and that this happens in nature also, is proved by the fact that Linneons which intercross in nature are known from almost all groups of animals and plants. Of animals we will mention of the class of the *Echinodermata*: Echinus esculentus and E. acutus in the neighbourhood of Plymouth (cf. Shearer, de Morgan and Fuchs Phil. Frans act. Royal Society B 204 pp.- 255—362), of the Class of the *Vermes*: hybrids between Ascaris univalens and A. bivalens with three chromoso-

---

[1]) This copulation has since proved to be fertile, but the young ones died soon after birth.

mes by Herla and Zoja (cf. Biol. CBl. 1912 p. 718), of *Arthropoda* it is well known, that in the group of the insects such hybrids are common. Among *Mollusca* Helix hortensis and H. nemoralis frequently cross, while among *Pisces* it is well known that f. i. in the ri- ver IJssel in Holland every hoal of the big nets called „blessings" (not from the fish's standpoint!) contains hybrids between the genera Leuciscus and Blicca, while hybrids between Blicca and Abramis and be- tween different Linneons of Leuciscus are common in many rivers. It appears that generally hybrids be- tween different Linneons of fishes are by no means rare (cf. Claus). Among *Amphibia* natural hybrids be- tween different Linneons of Triton are known (cf. Poll. Biol. CBl. 1909 p. 30) and also between different Lin- neons of frogs (cf. Boulenger). Among *Aves* hybrids are especially common between different Linneons, yes even between different genera f. i. of ducks, even such between individuals belonging to different families viz between Penelope (fam. Cracidae) and Phasianus colchicus (cf. Haecker p. 211) but this one not in nature. In Sweden Birkhahn and Auerhenne cross frequently, probably because the Auerhähne are decimated by shooting (cf. Naumann Naturgesch. der Vögel Mitteleuropas. Vol. 6 p. 106 Anm. 13.108. Among *Mammalia*, Darwin says in the Origin p. 224, that he has reason to believe that the hybrids from Cervulus vaginalis and Reevisii are perfectly fertile. So are the Indian humped and common cattle when crossed.

Among plants, hybrids between different Linneons

are quite common, and certain peculiarities in flower
structure or behaviour f. i. dichogamy even favor such
crossfertilisation among Linneons. An exemple of this
is furnished by the willows.

I will here quote what Kerner's Pflanzenleben says
about this (Vol. II p. 311)

„Auch die zweihäusigen Pflanzen sind in der Mehr-
zahl noch proterogyn. In den ausgedehnten Weiden-
beständen an den Ufern unserer Flüsse sieht man bis-
weilen einzelne Arten durch Tausende von Sträuchern
vertreten. Ein Teil derselben trägt Pollenblüten, der
andere Fruchtblüten. Sie wachsen auf demselben Bo-
den, sind in gleicher Weise der Besonnung ausgesetzt
und werden van denselben Luftströmungen bestrichen,
und trotz dieser gleichen aüssern Einflüssen eilen die
Stöcke mit Fruchtblüten ihren Nachbarn mit Pollenblüten deutlich voraus. Die Narben der Salix amyg-
dalina sind schon 2—3 Tage hindurch belegungsfähig
und dennoch hat sich weit und breit noch keine einzige
Anthere dieser Weidenart geöffnet. Dasselbe gilt van
der Purpurweide, der Korbweide, der Bruchweide
etc. Da das ungleichzeitige Eintreffen der Reife der
zweierlei Geschlechter einer Art eine Einrichtung ist,
welche bei den meisten, ja vielleicht bei allen Pflanzen
vorkommt, so kann auch nicht angenommen werden,
dass dieser Einrichtung gar keine Bedeutung zukomt. Ich
will es nun versuchen die Bedeutung der Dichogamie
zu erklären, und lade den Leser ein mit mir zunächst
eines der Weidengebiete zu betreten welches in vorher-
gehendem kurz geschildert wurde. Die Purpurweide be-
ginnt grade zu blühen. Die Fruchtblüten derselben

zeigen bereits belegungsfähige Narben, aber die Pollenblüten sind noch in der Entwicklung zurück und es ist
noch keine einzige Anthere derselben geöffnet. Dage-
gen stehen die Pollenblüten an der Korbweide (Salix
viminalis) welche untermischt mit der Purperweide im
selben Bestande wächst, auf dem Hohepunkt der
Entwicklung. Pollen der Korbweide ist in Hülle und
Fülle zu haben. Durch den Duft und die Farbe der
Blütenkätzchen angelockt, haben sich zahlreiche Bie-
nen eingestellt, schwirren von Strauch zu Strauch,
saugen Honig und sammeln Pollen. Sie sind bei dieser
Arbeit nicht wählerisch und beschränken sich nicht auf
eine einzige Art, sondern fliegen ebenso gerne zur Purpurweide wie zur Korbweide.... Indem aber die Narben der Purpurweide mit dem Pollen der Korbweide
belegt wurden, hat eine zweiartige Kreuzung stattge-
funden. Erst zwei oder drei Tage später kann eine einartige Kreuzung stattfinden, denn nun haben sich
auch aus den Pollenblüten der Purpurweide die Anthe-
ren hervorgeschoben .... *Bei Beginn des Blühens ist also*
bei der genannten Weide, infolge der Dichogamie nur
eine zweiartige, später erst eine einartige Kreuzung
möglich.

At the commencement of flowering, the Dichogamy of
the willow mentioned above, allows only crossing with
individuals of another Linneon, while later only fertili-
sation by members of the same Linneon becomes possible.

Of course it is indifferent in principle, whether the
pollen is carried by the wind or by insects, but as the
pollen, in order to be transferable by wind, must possess certain properties and the flowers which produce

such pollen also (in the first place to cover the great loss of such pollen during transportation, they must produce it in very large quantities) crossfertilisation between many Linneons of plants must have been almost impossible before insects came into existence.

The birth of insects consequently offered new possibilities of crossing and consequently of the birth of new species.

Darwin clearly perceived this, as results from a letter he wrote to HOOKER on Aug. 6th 1881 (Life and Letters III p. 248).

„Nothing is more extraordinary in the history of the „vegetable kingdom as it seems to me, than the *appa-* „*rently* very sudden or abrupt development of the „higher plants.... Hence I was greatly interested by a „view which Saporta propounded to me a few years „ago.... viz, that as soon as flower-frequenting in- „sects mere developed, during the latter part of the „secondary period, an enormous impulse was given to „the development of the higher plants by crossfertili- „sation being thus suddenly formed.''

Of course the influence of dichogamy remains the same in favoring crossfertilisation of Linneons whether the wind or insects are the transporters of the pollen, and so it is quite correct that Kerner makes the general statement, that dichogamy favours bi-specific crossing especially at the beginning and at the end of the flowering period of all plants possessing this peculiarity (cf. l. c. II p. 315).

That this hybridization of Linneons is by no means of rare occurrence, KERNER, who was the first to recognize the great importance of hybrids for evolution shows on p. 570, where he states that the number of such hybrids growing wild in Europe, which has become known in the last 40 years preceding the writing of his book, can safely be estimated at 1000 and that the 41 Linneons of Coniferae growing in Europe have produced no less than 7 hybrids.

When we consider further, that not only different Linneons but also different genera intercross f. i. Secale and Triticum, Triticum and Aegilops etc. and that up to the present, we know with certainty of no other way of the formation of new species and new Linneons than as a result of crossing, we may, I think, accept safely that the underlying cause of the diversity of the different types which people the earth, the underlying cause of evolution, at least as far as diploid organisms is concerned, is hybridization.

It draws a more complete paralell between the origin of individuals and that of species, on which Darwin insisted already, than his theory did, because according to his conception, species had but one parent, the varying ancestral species, while individuals had two parents; while according to our view, species as well as individuals have two parents, the first the two parental species from whose cross the new species arose, the latter their father and mother.

CHAPTER XII.

THE EFFECT OF CROSSING LINNEONS.

On talking with different scientific friends about the views which have gradually developped in my mind about crossing being the origin of species, I have frequently heard the objection — which is pretty generally offered against all theories of evolution — that the effect is inconsiderable, compared with the changes wanted to explain the origin of such different groups, say as fishes, reptiles and mammals.

I will return to this question in the next chapter, limiting myself in the present one, to show what the effect of an actually executed cross between individuals belonging to different Linneons has been.

It is impossible to treat here of all the numerous forms which result from such a cross; for this I must refer the reader to a book by the author, on the hybrids between different Linneons of Antirrhinum which would have appeared long ago, if the war had not inferfered with the making of the plates.

But as all that is necessary here, is to show the great diversity which can be the result of a cross, I can suffice by restating, what I said at the IVe Conférence de Génétique in Paris in 1911 and by reproducing some photographs as illustration.

In 1910 Professor BAUR of Berlin succeeded in crossing certain species of Antirrhinum, the hybrids of which proved to be fully fertile. The seeds of two of these hybrid-combinations he very kindly gave to me for further investigation.

One of these combinations was Antirrhinum glutinosum crossfertilised with a peloric form of Antirrhinum majus.

As Antirrhinum glutinosum contains several types, all of which are completely selfsterile, the F_1 generation obtained was somewhat polymorphous, though not to a considerable extent. On the whole, it was fairly intermediate between the two parent species.

The F_2 obtained from the F_1 plants was exceedingly polymorphous; one of the self-fertilised F_1 plants gave 255 children, not two of which were alike. They differed in a large number of characters, such as colour, form of flowers, habit of growth, leaf-characters, hoariness, self-fertility, resistance to draught and frost etc. Zygomorphous and peloric flowers were always present, and it appears that the segregation was a complicated mendelian one.

The most interesting results perhaps were obtained from the cross A. glutinosum by a red zygomorphous form of A. majus.

In the F_2 of this cross, several remarkable forms occurred, in one f. i. the sepals were coloured and petaloid, another showed several spur-like excressences at the lower lip of the flower and some had flowers astonishingly different from those of the parent-species, resembling more a *Rhinanthus* than an *Antirrhinum* and of a type entirely unknown hitherto within this latter genus, as the photographs here reproduced show.

The first photograph shows at the top, indicated as P¹ and P², the two parents crossed.

P¹ is the flower of Antirrhinum glutinosum, which is white, except that it has pink striae on the upperlip.

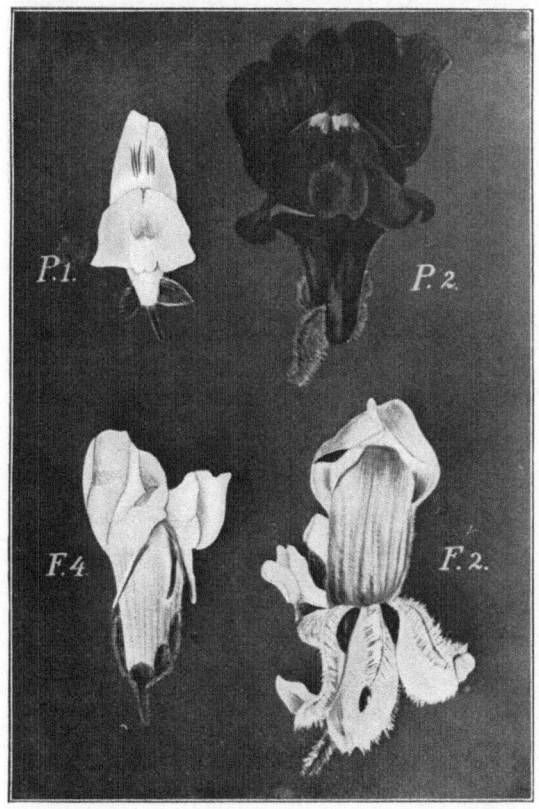

The flower indicated by P², is the dark red flower of Antirrhinum majus with which the former was crossed. Of the two other flowers on this photograph, the one which appeared in the fourth selfed generation (F₄), was pure ivory and characterized by the possession of spur-like organs at the lower lip, while the one which appeared in F_2 was light pink, and showed beautiful petaloid sepals of the same color but in a lighter shade.

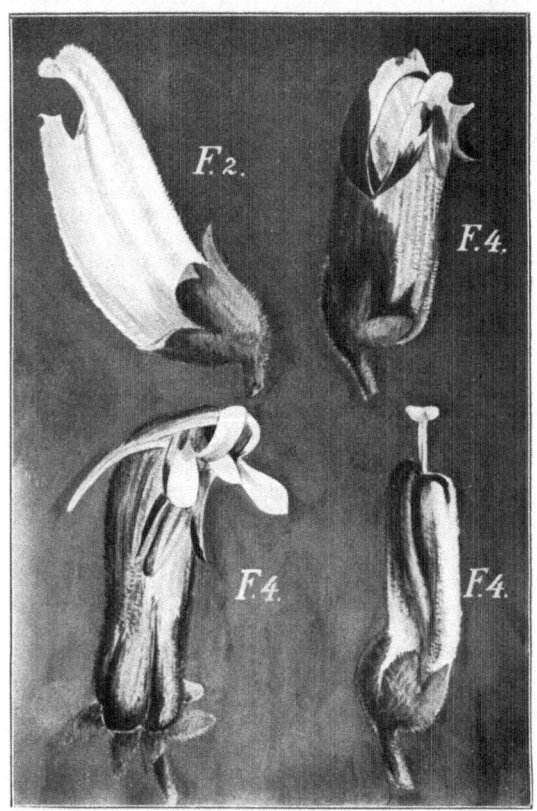

The second photograph shows one flower of a plant of the 2d generation, indicated by F_2, which was flesh-colored and resembled the flower of a Rhinanthus, the other flowers, all belonging to the 4th generation (F_4), were magenta and showed shapes very different indeed also, from the parent-species crossed.

It seems to me, that this little suffises to show that very unexpected results can arise from a cross, forms so different from the Linneons crossed, that no systematist could possibly say to which of these Linneons they belonged, yes so different, that if their origin were not known, they might easily be referred to different genera.

CHAPTER XIII.

THE ORIGIN OF THE GREAT CLASSES.

The origin of the great classes of the vegetable and animal kingdom is a historical problem in which geology has the first word.

The most important facts geology teaches us are:

1stly. that the origin of the different classes lies very far back.

2dly. that a new class appears suddenly with a great many, mostly highly differentiated, different forms.

3dly. that the further we get away from the birth of the class, the more the number of different forms diminishes, and the more „reduced" these become.

We will illustrate these important facts by some examples.

SCOTT says (p. 6. of the Introduction to his study in fossil Botany 1908) about the very ancient origin of the great divisions:

„There are probably no biologists left now who „oppose in toto the doctrine of evolution, but if there „were, they might draw a telling, though fallacious „argument from the high organization of the Devonian „flora".

The sudden appearance in a large number of different forms, of such groups as the Reptils, Cycadophytes

and Angiosperms is generally known; also their gradual diminution in types (except in the latter group,
which is yet flourishing) until but a few remain, or total
extinction follows.

If we compare the Cycadophytes of Mesozoic times,
with their highly developed Bennettitales, with the few
Cycadaceae of to day, or the Lycopodiales of the Coal-
measures, with their giant Lepidodendrons, with the
miserable Lycopods of to day, or the Equisetales of the
Coal-measures, with their numerous giant Calamites,
with the few horse-tails of the present time, or the fossil
Reptils and Ammonites of the Jura with the recent
ones, the exactness of the facts mentioned above, sub 2
and 3, is apparent.

*Now all this coincides splendidly with what we see
happening after a cross between two differentLinneons,*viz.
1stly.   appearance, among the progeny of the hybrids,
         of a very large number of different forms.
2dly.   gradual decrease of that large number of diffe-
         rent forms until but a few remain within each
         Linneon.

Soon after the formation of a large number of different forms from a cross, these continue to intercross,
leading thus to a constant increase of new types, but
gradually this intercrossing ceases, either by isolation,
sterility or by any other obstacles among which aversion
may play a considerable rôle, and so smaller groups, Linneons, find themselves together as communities to which
intercrossing, if such continues, is limited. If it does not
continue, mere segregation will rapidly cause an increase of the homozygotes in proportion to the heterozy-

gotes, and these homozygotes will be decimated in the struggle for life; while if intercrossing continues, selection of the dominant type which seems to be the rule, causes apparent uniformity within such Linneons, as in the case of the grey rabbits.

The result is, that from the large number of types arisen from the cross, finally but a few survive.

Exactly what we see *en grand* after the appearance of a new Class of organisms.

Within each Linneon the number of smaller units, of Jordanons, decreases also by selection, — which spells: extermination (of the less fitted types) — until finally but a few or even but one survives.

If one survives only within a self-fertilising Linneon, its „variability" is at an end, and therefore it must sooner or later die out, and the same must happen if, within a Linneon, while intercrossing of the different types it contains continues, selection leads to the survival of the recessive type, in stead of to that of the dominant one, because then its „variability" is at an end also, and extinction must follow.

That exclusive survival of recessives is one of the chief-reasons of extinction, gains support from the fact that the recent forms of ancient groups are all weak ones, compared to the older ones, and experience goes far to prove that the recessive types are usually the weaker ones.

Crossing therefore is the cause of the origin of new types, heredity perpetuates them, selection is the cause — not of their origin as was formerly supposed — but *of their extinction.*

Selection, if not interfered with by crossing, inevitably ends with extinction because selection can act through extermination only, and by thus diminishing the number of differently constituted types, decreases the possibility of crossing and thus the formation of new types, which formation not only means the origin of new species, but also the possibility of adaptation, as the only way in which a Linneon can adapt itself to new circumstances, is a putting into the world of new types.

All this agrees perfectly with what we saw that happens, in the appearance and gradual extinction of new classes of plants or animals, so that we are justified to conclude:

Crossing was the origin of the new classes; selection, the result of extermination by the struggle for life, the cause of their gradual extinction.

Such extinction of classes must proceed continuously, until a happy meeting between two sufficiently differently constituted gametes, causes the origin of a new class.

We will therefore have periods of new-formations of classes and periods of gradual extinction of such classes.

In the present time we live in a period, in which the extinction of many classes is almost completed, and in which no new classes are formed.

This is easily demonstrated for the vascular plants by a glance at Scott's system:

Sphenopsida
{
Equisetales.
Pseudoborniales.
Sphenophyllales.
Psilotales.
}

Lycopsida   Lycopodiales.

Pteropsida
$\begin{cases} \text{Filicales.} \\ \text{Pteridospermeae.} \\ \text{Gymnospermeae.} \\ \text{Angiospermeae.} \end{cases}$

Of these three great divisions the two first are extinct, with the exception of a few Equisetaceae, Psilotaceae and Lycopadiaceae; of the Pteropsida the Pteridospermeae are extinct, Gymnospermeae on the way to extinction, so that only the Filicales and the Angiosperms remain, of which the latter alone, are flourishing.

A formation of new classes is not in action at the present moment, so that it is illegitimate to claim, that one who wants to explain evolution, must demonstrate how such a formation of new classes goes on.

One can 't demonstrate something which does not happen, at the time one is living; we must conclude to the way of the origin of new classes by analogy, and we have done so, by comparing it with what happens after crossing two Linneons. What happens there, is demonstrable and we have demonstrated it.

Whether new classes will in future be formed again on our globe, or whether we assist at the last trial of life to maintain itself on our globe, in other words whether we are living at the beginning of the period of extinction of the latest born classes, we of course know not.

CHAPTER XIV.

THE QUESTION OF RELATIONSHIP.

Relationship among different groups of animals and plants has generally been considered to be of the nature of bloodrelationship.

So Scott says: „In these days most of us, when we speak of relationship among organisms, mean to imply „a real affinity, that is to say a bloodrelationship."

And a little further on he specifies:

„Existing organisms are related to each other more „or less as brothers or cousins are related."

Now the question is: *are* brothers related?

As all men are hybrids, the bloodrelationship beween members of the same family has been very much exagerated; as a matter of fact the constitutional similarity — which, though wrongly, is meant when one speaks of bloodrelationship — can be very slight between brothers and sisters, yes can be much less even, than between individuals of much more remote parentage.

As similarity of the individuals depends — abstraction made from external influences — on the constitution of the gametes which formed these individuals, this similarity will reach its maximum, e. g. become identity, in case gametes of identical constitution mate, irrespective of the source of these gametes, while it is perfectly indifferent, whether such gametes of identical constitution derive from hybrids or from specifically pure individuals.

Consequently, we may confidently expect to meet occasionally very similar human individuals, who are not al all related in the usual sense, and we all know that this happens so frequently, that it is even said that each man has his duplicate somewhere.

What value then have we to ascribe to the bloodtest of UHLENHUTH and NUTTALL?

No other, than revealing, perhaps, similarity of constitution, which is quite another thing than relationship.

As little as the result of testing for *Ca* in chemistry, revealing the presence of this element in the stalactites of a grotto and in the substance of our bones, reveals the derivation of the calcium in the stalactites and in our bones from the same source — much less the derivation of the stalactites and of ourselves from the same source — as little the results of the bloodtest reveal relationship. *Yet, in a sense, it is correct, that species are related to one another in a similar way as human brothers or cousins are, viz in as much, as this expresses the fact that both are segregates from a former cross.*

In the case of brothers and cousins, we can trace their origin if there is a family register, if no such register of births exists, we can not; so that, if species are really „related" in a similar way, it is, in the absence of species-birth registers, a priory improbable that we shall ever be able to reconstruct their phylogeny.

CHAPTER XV.

THE QUESTION OF PHYLOGENY.

As long as one accepted an uninterrupted progression say from amoeba to man, it was justifiable to deduce from the degree of similarity of a given form to the amoeba, or to man, the place which this form ought to occupy on the ladder leading from the amoeba to man.

Of course, the stronger the belief in orthogenetic evolution, the greater the degree of certainty with which the place of a given form in phylogeny could be determined.

Some such degree of certainty could be attained in Lamarck's evolutionary scheme, because this supposed all progress to come about by response to necessity, so that, if it was for-ordained that the amoeba should better its constitution so as to reach that of man, there was reason to suppose a rectilinear succession of types also.

Nägeli's Vervollkommnungstrieb allows a similar degree of certainty. This idea of a rectilinear progression was given up by Darwin, who saw the cause of progression in the selection of the best of a number of allsided variations, comparing the road along which phylogeny proceeded, to a branched tree, in which the position, to be assigned to a given individual, became very doubtfull.

Yet, as long as one believed similarity to be a measure for the degree of relationship, optimistic natures could continue to believe that arrangement of the different types according to their degree of similarity, would give a picture — be it a blurred one — of phylogeny.

Now we know that similarity, neither phenotypical nor genotypical similarity, is a criterium for relationship.

Sisters can be more dissimilar than two girls of quite different extraction, and consequently similarity gives us no measure whatever of relationship.

And as we know furthermore, that crossing has occurred and does occur daily, that it produces many different types, from which but a few survive, after having recrossed in alle directions, we know that evolution proceeds forwards, sidewards and backwards, along the meshes of a net, so that it is absolutely hopeless to choose out of the many ways, in which one can draw a broken line on such a netting, the one along which evolution has proceeded.

Phylogeny e. g. reconstruction of what has happened in the past, is no science but a product of phantastic speculations which can be held but little in check by the geological record, on account of the incompleteness of the latter.

Those, who know that I have spent a considerable part of my life in efforts to trace the phylogeny of the vegetable kingdom, will know, that this is not written down lightly; nobody cares to destroy his own efforts.

Every author, who cares to be taken as a man of

science, has however to expose his own errors in the first place.

Adhering to ideas venerable by age, irrespective of the possibility of proof, is the prerogative of religious creeds; science ruthlessly destroys what it no longer recognizes as well established.

There is something very dear to the human heart in sticking to venerable traditions; it is even questionable whether the upholding of illusions is not better worth our while than the restless pursuit of the truth, which forces us to destroy to day, what we built up yesterday, but this is a philosophical question which does not concern us here, science demands the total eradication of convictions as soon as these are recognised to be ill-founded.

Happily we but rarely err completely, and though Phylogeny is untraceable and consequently, as said above, a product of phantastic speculations, yet the basic truth of it remains unimpaired, which is so well expressed by Darwin in the conslusions to his origin:

„*As all the living forms of life are the lineal descen-* „*dants of those which lived long before the Cambrian* „*epoch, we may feel certain that the ordinary succession* „*by generation has never once been broken.*"

In this sense Phylogeny is a fact; it are our reconstructions of the way in which it proceeds (which we use to call Phylogeny) only, which are phantastic.

The recognition of this is, to my mind, a fact added to our stock of knowledge, be it, in a sense, a negative one.

———

CHAPTER XVI.

HOMOLOGOUS PARTS AND RUDIMENTARY ORGANS.

What is meant by homology?

Darwin expresses it as follows:

„The members of the same class, independently of their „habit of life resemble each other in the general plan „of their organisation. This resemblance is often „expressed by the term „unity of type" or by saying „that the several parts and organs in the different „species of the class are homologous"

Hence homologous parts are parts occupying corresponding positions in the general groundplan.

Therefore one is tempted to consider the corollae of two flowers to be homologous, however different, while bracts, however corolla-like, are certainly not homologous with the petals which form the corolla, but only analoga of these.

Now the determination of what is homologous and what is analogous, consequently depends on the certainty with which we can determine the position of a part in the general groundplan, and as the name „general" groundplan already indicates some uncertainty in this respect, it is no wonder that one meets in the literature with endless discussions as to which parts are homologous and which analogous. It was to call attention to this fact, that I said above, that we are *tempted* to consider all corollae to be homolo-

gous, because we all know that it is exceedingly difficult to say, in a particular case, whether the place, occupied by the corolla in the general groundplan, is the
rightfull place of the stamens, that of the calyx or e-
ven that between the outer whorl of stamens and the
calyx.

In the first case, such a corolla would be the homologon of the stamens, in the second that of the calyx,
and in the third place, it would have no homologon at
all, but be a new formation, an intercalation in the
general groundplan.

The trouble in questions, like those of a general
groundplan, and of the character, the type, the essence
(wesentliche charactere) of a Linnean species, is that
we are unable to grasp all the characters of a group or
even of an individual, and so are led, unconsciously,
to consider the most conspicuous ones, as the essential ones. Gradually we find out, that among these
most conspicuous characters some are common to
more members of the group than others and as men are
great believers in the principle of majority (does not
a superior minister when he finds himself unable to
solve a difficult problem, appoint a commission of
M. P's, probably each less capable than himself, and
accept their majority-report as decisive?) we are led to
take these characters in the majority to be the best, to
be *the* most essential, to constitute in one word *the*
groundplan. It ought to be plain to us, that this is
nonsense, that we have thus succeeded in construc-
ting a common groundplan only, by divesting the
plans of the different individuals from all what made

them different, and that we can do this just as well with members of very different classes as with members of the same class, by trimming their differences a little more yet.

It is easy, in this way, to construct a general ground-plan even between a sea-urchin and a hedgehog, by abstraction of all the differences they show, until we have nothing left than the one point they have in common: the covering with spines and then claim that the groundplan of Echinus and Erinaceus al-lows us, to unite them to the group of the Spiniferae, which is not very much worse than the group of the Chordata as defined in the Handwörterbuch of the Naturwissenschaften II p. 623 as: „die durch den Be-sitz einer Rückensaite, Chorda dorsalis, ausgezeichne-ten Tiere: Tunicaten, Amphioxus und Vertebraten."

Man's mind is a curious one, because, after having thus divested the different individuals he sees in na-ture, of all their differences, he assigns such immense importance to the shabby rest thus obtained, that he proceeds to explain the differences (he has just succeeded in explaining away) by speculations about the causes which could have changed this common rest to all the different structures which are actually presented by the different organisms, *forgetting enti-rely, that there is not the slightest reason to consider this rest as ever having been transformed.*

It is always the same mistake: looking for the (non existing) single origin of different types.

So one reaches all kinds of attractive, but quite un-founded, conclusions as f. i. that the flapper of a seal is a metamorphosed hindleg of a landanimal, which conclusion is about as well founded as that the door of my house is a metamorphosis of the door of my neighbour's.

This may have happened, but *need not* have happened; my door may have been made from that of my neighbour's, by taking off something here and adding perhaps something there, so as to fit my house, but it may just as well, and much more probably has been, constructed anew.

So why should we accept the flapper of a seal to have been formed by the changing of the hindleg of a land-animal? We know, that from a cross not one type, but several types arise, so that there is no reason whatever to consider all these different types to be derived from the one with the simplest constitution; we know on the contrary that it are just the comparatively simplest types, the recessive ones, which are entirely unable to give rise directly to other ones.

A type giving rise to a large number of forms must on the contrary be a *complicated* one, allowing segregation.

I hear it objected, that I am proceeding to do the same I reproach to those I criticise, viz that I am on the way to explain away the differences which *do* exist between different classes.

This is not my purpose; I only claim that the „groundplan" is but a very general one, that we are not justified to consider the simplified „groundplan", obtained by cutting off all differences in structure existing between the different organisms, as primitive, as the one possessed by the common ancestor of those organisms, and that consequently we have no reason to suppose that each of the parts of the different organisms had already a position in that ancestral organism, through which we can spot the homologies of a given part of an organism, now existing.

It remains perfectly true however, that the members of not too large a group resemble each other in their general plan of organization.

Now this is stating a fact of the same kind as when we say that the different forms resulting from a cross between two Linneons resemble each other in the general groundplan of their organisation. The cause must lie — in the latter case at least, in the former we do not even know whether the members of the group belong genetically together — in the constitution of the gametes which united to form the hybrid, which initiated the group of new species, and as long as we know nothing of the constitution of these gametes, we can explain, as little how this comes about, as we can explain why one chemical substance cristallizes in the one, another in another form.

Now one hears it often said, that there is a basic difference between the structure of new Linneons, arising from the cross of two preexisting ones, and between the structure of a new class of organisms, supposedly also arisen from the cross of two individuals — be it from two more different ones — in as much as the general plan of the new Linneons is the same as that of the Linneons crossed, while the general plan of the new class must, on the contrary, be different from that of the class to which the crossed individuals belonged.

But this statement is not correct. The difference is but one of degree. A cross between different Linneons also gives rise to differently constituted types and to consider these to be of the same general plan as the Linneons crossed, is again only possible by abstracting from the differences.

If we cross f. i. two white Sweetpeas — even such belonging to the same Linneon — we may obtain colored types, some of which have exclusively colored descendants although they may segregate into many *differently* colored types.

Now a group of such colored types has a different „groundplan" from the uncolored types from whose cross they arose; we are even able to say in what these groundplans differ. They differ in so far as the colorless types crossed, possess only one substance of a certain class, either a chromogen or an oxydase, while the colored groups, resulting from their union possess both substances in their „groundplan."

Now, mutatis mutandis, the skelettal groundplan of the vertebrates may have arisen by the crossing of two invertebrates, each of which possessed some of the substances necessary to form a skeleton, but lacked some of the others, which combined by crossing these two types.

And just as in the colored Sweetpeas mentioned above, the groundplan „color" remains, but is different in the different segregates from some of these colored types, so the groundplan of the skeleton may remain in some of the descendants of the crossed invertebrates — which descendants we then unite to the class of the vertebrates — but be different in the segregates arisen from them.

So the different forms of extremities we meet with, in different vertebrates, can be explainèd in a similar way as the different forms and colors of flowers, viz by segregation, and those existing now, are the rests of probably many others which have existed formerly, as becomes evident upon comparison of the extremities of the now existing reptils with those of former geological periods.

The principle of selection holds good in as far as it shows the *result* of the extermination by the struggle for life; it shows us which forms could resist this, but it is no principle which explains the origin of certain types; selection also spells extermination, the types last to be exterminated, obtaining the *epitheton ornans*: selected types.

Consequently, we have to drop the idea of homology in the sense of parts, occupying corresponding positions in the *ancestral* groundplan, but may continue to use it, for convenience sake, *cum grano salis*, as in-indicating corresponding positions in the *general* groundplan, if we only keep in mind that such mapping out has no other significance than as a *pons asinorum* for our memory and never allows us, with certainty, to distinguish between analogous and homologous parts or between these and new formations, as is shown by the example of the corollae, mentioned at the beginning of this chapter.

The question of rudimentary organs is of no special significance to us; it only states the fact of the existence of organs which evidently lack something to become completely developed, just as the white flowered sweetpeas lack a substance, the presence of which would allow them to develop color, and can therefore be described as to be rudimentary colored.

When we look back on what has been said, it results that it is very difficult to say in a given case whether a form is primitive in its class or reduced, in other words whether it is near to the original hybrid which initiated the group, or further removed from it, because as we have seen, a F_2 generation already can contain types of very different constitution, so that the designation primitive or derived, looses much of its meaning.

In a very general way however, we can say that there is evidence of the non-primitiveness of most of the simpler types, in this sense that the simpler types usually are recessive segregates and consequently unable to give rise to new forms, unless crossed.

It certainly speaks volumes for the genius of Charles Darwin, that although differing completely from him in my opinion as to the way in which evolution takes place, I can nevertheless conclude this chapter with the very words with which he concluded his corresponding chapter:

„Finally the several classes of facts which have been „considered in this chapter, seem to me to proclaim so „plainly, that the innumerable species, genera and

„families with which this world is peopled are all des-
„cended, each within its own class or group from com-
„mon parents, and have all been modified in the course
„of descent, that I should without hesitation adopt
„this view, even if it were unsupported by other facts
„or arguments."

———————

CHAPTER XVII.

MIGRATION.

That the types put into the world by a cross, do not always remain at the spot of their birth, but migrate, is a generally known fact.

That they can do so in an astonishingly extensive way is known also, and that it is done sometimes — frequently perhaps even — by the intermediancy of man, such as by travelling man himself, by his beasts of burden or by his trains or vessels, of course, abstracts not a *iota* from the fact of such migration existing, but only justifies the conclusion that this kind of transport cannot have existed in times before man appeared on our globe.

It does not prove of course, that migration was less extensive in former periods than now, because other ways of migration then existed than now, f. i. land-connections allowing intercourse between continents now separated, more rainfall preventing the erection of such formidable bars against migration as the Sahara-desert, higher temperature of the sea, allowing animals and plants of the aequator to migrate further towards the north and south, than is now possible, less interference of man, by non-extermination of types considered a nuisance by him etc. etc. So that we have no measure for the relative degree of migration going on in different geological periods; all we can say is that migration formerly existed as it does now.

Yet migration has been exagerated in one respect, in as much, as here also, as in former theories of evolution, the dogma of the single origin reigned supreme.

So, if Primula acaulis is found in England and in the Tirolese Alps, it is concluded that it must necessarily have originated either in England or in the Tirolese Alps or on a single third spot from where it has migrated to the different countries it new occupies.

Now this creed was the consequence of the supposition that species arose by selection of favorable variations, and that, what was favorable at the one spot, could not be expected to be favorable — at least not to the same degree — at another spot, so that the result at different spots could not be the same e. g. that the production of identical species at different spots, was well nigh impossible. This objection — in itself not well tenable because during this selective process the ancestral species itself does not remain stationary, but also migrates — of course falls away entirely if species are not the result of a selective accumulation of favorable variations, but are born, ready made, as a result of a cross and left to try to find a place fit to support them, or... perish.

If we dwell a moment on the fact, that identical gametes, irrespective of their source, must give the same kind of zygote, irrespective of the spot where the mating occured, it follows from the theory of the origin of new types by crossing, that identical types can originate at different spots, that species can arise polytopically.

This has long been conceded for „varieties"; it was BRIQUET who first claimed its validity for species and he is in my opinion perfectly right.

This way of a polytopic origin of identical types, makes it possible to divest migration of many very uncertain attempts to explain the present occurrence of identical types in very far distant places, but does not tend to make the present geographical distribution of animals and plants more easy of explanation in detail than it formerly was.

Geographical distribution is, just like evolution in the past, a historical problem which can never be reconstructed completely, but from which we can derive no more than the very general underlying principles.

CHAPTER XVIII.

GEOLOGY AND THE CONSTANCY OF SPECIES.

The selection-theory, as well as the theories based on inheritance of acquired characters, must suppose that variability, will it have any effect, must occur more or less continuously, that in other words organisms can't survive very long without showing variation. It is even inconceivable, on the ground of these theories, that the groups considered by them to be primitive, say like the Flagellates, should yet remain in existence, notwithstanding all the changes which the earth has undergone must have been inductive to their varying. This applies even, though in a lesser degree, to the mutation theory, the only conception able to explain the unchanged continuation of a homozygotic type through long ages, is the conception that a species is constant, unless it happens to cross with another one.

Now what does the geoloigcal record teach us to this effect? Let us quote here what GRAND' EUEY than whom no one, has greater experience in this matter, says:

„*Un fait notoire domine tous les autres, la permanence* „*des espèces durant la majorité ou la presque totalité de leur* „*existence*".

Venu il y a 25 ans à St. Etienne avec l'idée contraire que les espèces ont varié d'une manière continue, D.

STUR me conseilla de m'en assurer sur le terrain. Au lieu de cela, dans des dépôts interrompus, qui auraient conservé les formes gradiées d'espèces variables je n'ai rien rencontré que les débris d'espèces constantes; à l'appui de ce dire il me serait facile de citer plus de dix espèces communes aux deux grands bassins houilliers français, plus de dix espèces immuables de la base au sommet du bassin de la Loire et plus de dix autres transversant sans changer la moitié supérieure de ce bassin"

If we consider further that M. Grand' Eury's experience shows that „les genres les plus naturelles commencent par d'espèces peu distinctes et mélangées et une fois fixées se séparent et ne changent plus", we can only say that this agrees astonishingly well with what happens after a cross, and that thus geology gives strong support to the conception that crossing is the underlying cause of the origin of new types.

CHAPTER XIX.

CONCLUSIONS FROM THE BEHAVIOUR OF DIPLOID ORGANISMS.

New forms arise as the result of a cross; they can gradually become specifically pure by segregation *if self-fertilization* prevails.

Once pure, they perpetuate themselves by heredity and constitute a species.

Several of such species are united by systematists to Linneons. A Linneon of this kind *e. g. a Linneon consisting of habitual selffertilizers*, gradually looses species by extermination through the struggle for life, which process may result in the survival of one species within such a Linneon only, which species is then called selected.

Selection therefore spells: extermination.

Such a Linneon, reduced to one species, becomes synonymous with species and can withstand changes of conditions only, by non-transmittable plasticity, which is frequently called adaptability.

The usual idea that a species can survive by adapting itself to changed conditions by transmittable variability, is not only wrong, because such transmittable variability does not exist, but also because the result of such a process would be the creation of new species and consequently not assure the survival of the old one, but its replacement by others.

A Linneon of the class here treated, once having become monospecific, must sooner or later be exterminated without leaving any descendant, unless it crosses with a form belonging to another Linneon. Such a progeny of course is not identical with the parent-species, but gives rise to new species.

The conclusion therefore is: *species arise by crossing, perpetuate themselves by heredity and are gradually exterminated by the struggle for life, those last exterminated obtaining the epitheton ornans: selected ones.*

The result of a cross consequently *can* lead to the formation of new s p e c i e s but *need* not do so; it always *must* lead to the production of new f o r m s.

If however such new forms continue to intercross, no new species will ever be formed; usually however crossing is not so promiscuous as all that, and new species do arise by segregation; to such applies what has been said above.

Among the other new forms, promiscuous crossing sooner or later is limited also to intercrossing within certain groups, between which bars against intercrossing are erected by isolation, aversion or sterility.

Such smaller intercrossing communities are also united by systematists to Linneons.

They can resist untoward circumstances much better than Linneons, consisting of strict selffertilizers, because they have a far greater degree of plasticity through the fact, that the intercrossing of the forms within them, gives rise continually to the birth of new forms, offering new material, resistant to the exterminating effect of the struggle for life.

As long as the new forms, resulting from such intercrossing, do not transgress the border of the Linneon,
we can say, with justice, that such a Linneon maintains
itself by adaptation, although of course the adapted
Linneon is then internally changed e. g. is a group of
types of other constitutions, than the ones it previously
contained.

In such a Linneon also, the struggle for life gradually
plays havoc because, by extermination of certain types,
it gradually reduces the scope of crossing and consequently of adaptability. The struggle for life usually
results here, first in the selection of the dominant type
which can never proceed so far that pure dominants
only survive, because the hybrids with the dominant
characters enjoy the same advantages as the constitutionally pure dominants in the struggle for life
If therefore, the dominant form is selected, this is of
great advantage to such a Linneon, because then it
continues to contain the recessive forms — be it cryptomerously — and consequently can call on them in
case of emergency. As however such a call is hampered
by the fact, that such recessives can come into being
only, if two impure dominants happen to mate, even
the selection of the dominants must in the long run
lead to extinction. Such extinction will however come
about much quicker, if the struggle for life results in the
selection of the recessives, because then the Linneon
finally becomes really mono-specific and so must
perish sooner or later, exactly like the self-fertilising
Linneons treated off above.

The *conclusion* therefore is: *crossing of allogamous forms leads to the production of new forms, most of which sooner or later fall into separate non-intercrossing groups, each of which however consists of different intercrossing forms. Such groups are called Linneons.*

Such Linneons can adapt themselves to changed circumstances by giving birth to new forms by crossing the different types they contain. These new forms do not transgress the limits of the Linneon and are gradually also exterminated by the struggle for life, which ends in the selection, either of a dominant hybrid type ill fitted to give rise to new-forms, or of a recessive type entirely unable to form new forms.

Such Linneons consequently also, are gradually exterminated by the struggle for life, and here also the forms within it, last exterminated, obtain the epitheton ornans: selected ones.

We can express this shortly by saying:
Linneons arise by crossing and are gradually exterminated by the struggle for life, the last surviving ones obtaining the epitheton ornans: selected ones.

All this is probably parallelled, en grand, in the appearance and extinction of the great classes. A cross between two greatly differing gametes gives rise to a great diversity of new forms which we group into families, genera and Linneons, to all of which applies what has been said above of the Linneons: gradual dimunition of the opportunities of crossing and consequently of the birth of new types, able to withstand changed conditions. This must

finally lead to extinction of the classes also. There is
consequently periodicity in the production of new
types, no matter whether these are so different that we
put them into different classes, or so little different that
we put them into different Linneons only.

Such a period of production of new types is always
immediately followed by the beginning of extinction
through the struggle for life, which is withstood as long
as crossing remains possible, and becomes complete
sooner or later, after the last possibility of a cross has
disappeared.

Selection is only „une belle phrase" for extinction, the
forms last exterminated, being called the selected ones.

Adaptation has a double meaning: *adaptation of Linneons* to changed conditions by changing their internal
composition through the production of new types as the
result of crossing, which do not transgress the limits the
Linneon *and adaptation of individuals* by the plasticity
which every individual enjoys, within limits, to respond
to the call of new necessities.

*The vera causa of the production of new types conse-
quently is: crossing; the vera causa of their extinction:
the struggle for life; the selection resulting from the latter,
is by no means a revival, but is the sign of struggle of the
doomed.*

*The production of new classes can evidently only be
studied and demonstrated experimentally in a period of
production of such classes; in a period, as the present one,
we must be content with the demonstration of the origin of*

new Linneons, and must conclude by analogy to the way by which new classes originated.

I should like to finish this chapter by calling once more attention to one of the chief modern mistakes, which, according to my view, has been made in the interpretation of evolutionary facts. This is: that one has looked for the cause of the origin of the different types within a Linneon exclusively within the limits of that Linneon, and so has been led to conclude, in most cases, that the most common wild form within that Linneon, was the ancestral one.

So BATESON says, after crossing differently constituted white flowered types of the Linneon, Lathyrus odoratus:

„When F_1 was grown however, it was clear that here was „a remarkable opportunity of studyng a reversion „in colour, due to crossing, for these plants instead of „being white were purple, like the wild Sicilian plant „from which our cultivated sweet peas are descended.''

And in his presidential Australian address he says about this same point:

„In spite of repeated trials no one has yet succeeded „in crossing the sweet Pea with other leguminous „species. We know that early in its cultivated history „it produced at least two marked varieties which I can „only conceive as spontaneously arising, though no „doubt, the profusion of forms we now have, was made „by the crossing of those original varieties''.

Now why accept another origin e. g. spontaneously arising for these two „original'' varieties than for those arisen later?

My contention is that they were segregates from impure purples, indistinctible from pure purples, which
grew intermixed with the pure purples in Sicily, and
which revealed their impure nature after isolation in
domestication, just as my impure brown-yellow wallflowers segregate, if isolated and selfed: violets, golden-
yellows and whites.

*The real origin of the different types which we unite to a
Linneon lies not inside that Linneon — although part of
them arose from secondary intra-linneontic segregation
— but lies further back in the cross of two individuals,
belonging to different other pre-existing Linneons.*

HAPLOID ORGANISMS AND MUTATION.

For the investigation of the existence or non-existence of mutation, no group of plants offers better opportunities than mosses.

Bacteria are all too uncertain for such investigations on account of the possibility of contamination, especially if great care is not taken to start with absolute certaintly from a single cell. The delution-method, so frequently ressorted to, never gives *absolute* certainty in this respect.

Organisms with multinucleate cells are never safe objects because BURGEFF has shown, in his most interesting investigations on the results of crossing different types of Phycomyces that such cells may be heterocaryotic e. g. may contain nuclei of different constitutions, so that with such polyenergid organisms one can at best, obtain the same relative certainty for their specific purity, as is obtainable in the case of diploid organisms, but no greater certainty.

But mosses offer very much better opportunities for the final settlement of the vexed mutation-question.

As each mossplant arises from a single gamete, it can not be an inter-gametic hybrid as diploid organisms so frequently are. Leaving for the present the question of the possibility of intragametic hybrids aside, we can therefore say that, if it could be proved that a moss plant were able to produce more than one kind of gametes, mutation would have been proved in so far at least, as we would be justified to conclude, from such a result, that a monogametic organism can become polygametic without the direct interference of a cross.

Experiments with mosses, from which of course hybrid diploid generations can be obtained, viz hybrid capsules, are therefore highly advisable, also because they can throw light on the question of gametic purity, the nature of every separate gamete here becoming visible in the mossplant, which arises from it, while in diploid organisms we can see but the effect of the interaction of two gametes.

CHAPTER XXI.

DIPLOID ORGANISMS AND MUTATION.

We have given our reasons for being sceptical as to the existence of mutations.

This scepsis is partly based on fact, partly on circumstantial evidence. The fact is, that we possess no means to establish complete specific purity experimentally, and consequently are never sure of the purity of our material.

Certainty of purity however is a conditio sine qua non to obtain proof of the existence of mutation in living beings, just as chemical purity is a conditio sine qua non to obtain proof of the existence of mutability of the elements.

The circumstantial evidence is, that all so called mutants are recessives e. g. arise in material in which impure individuals are indistinctible at sight from pure ones, which makes it very probable that the aberrant individuals were no mutants at all, but segregates from heterozygotes, indistinctible from the pure dominants.

Circumstantial evidence always contains an element of uncertainty and the evidence from the recessives is no exception to the rule.

The fact that the „mutants" are always recessives is namely no *proof* that they are segregates.

If the presence- and absence-hypothesis holds good, they may really be mutants, arisen by a loss of a gen or factor, as a result of faulty inheritance e. g., after our present conceptions, as a result of accidental irregularities in karyokinesis, but even if this were the case, such an occurence would not materially assist the mutation theory because evolution by a process of repeated losses is inconceivable.

The mutation theory requires proof of the existence of progressive mutants e. g. proof that dominants can arise from recessives.

As we have seen, there is not a particle of proof for such an occurence. It need hardly be added, that it is not sufficient to prove the occurence of a dominant in a bed of recessives, but that it is necessary to prove that the dominant arose from a recessive.

The mere presence of a dominant in a bed of recessives, proves as little that it arose from a recessive as the presence of a cuckoo's egg in the nest of a sparrow proves that this egg arose from a sparrow's egg.

FINIS.